Springer Theses

Recognizing Outstanding Ph.D. Research

Aims and Scope

The series "Springer Theses" brings together a selection of the very best Ph.D. theses from around the world and across the physical sciences. Nominated and endorsed by two recognized specialists, each published volume has been selected for its scientific excellence and the high impact of its contents for the pertinent field of research. For greater accessibility to non-specialists, the published versions include an extended introduction, as well as a foreword by the student's supervisor explaining the special relevance of the work for the field. As a whole, the series will provide a valuable resource both for newcomers to the research fields described, and for other scientists seeking detailed background information on special questions. Finally, it provides an accredited documentation of the valuable contributions made by today's younger generation of scientists.

Theses are accepted into the series by invited nomination only and must fulfill all of the following criteria

- They must be written in good English.
- The topic should fall within the confines of Chemistry, Physics, Earth Sciences, Engineering and related interdisciplinary fields such as Materials, Nanoscience, Chemical Engineering, Complex Systems and Biophysics.
- The work reported in the thesis must represent a significant scientific advance.
- If the thesis includes previously published material, permission to reproduce this must be gained from the respective copyright holder.
- They must have been examined and passed during the 12 months prior to nomination.
- Each thesis should include a foreword by the supervisor outlining the significance of its content.
- The theses should have a clearly defined structure including an introduction accessible to scientists not expert in that particular field.

More information about this series at http://www.springer.com/series/8790

Vladimir Kulikovskiy

Neutrino Astrophysics with the ANTARES Telescope

Doctoral Thesis accepted by
the University of Genoa, Italy

Author
Dr. Vladimir Kulikovskiy
Physics Department
Università degli Studi di Genova
Genoa
Italy

Supervisor
Dr. Marco Anghinolfi
INFN Sezione di Genova
Genoa
Italy

ISSN 2190-5053 ISSN 2190-5061 (electronic)
Springer Theses
ISBN 978-3-319-38715-4 ISBN 978-3-319-20412-3 (eBook)
DOI 10.1007/978-3-319-20412-3

Springer Cham Heidelberg New York Dordrecht London

Softcover reprint of the hardcover 1st edition 2015

Printed on acid-free paper

Springer International Publishing AG Switzerland is part of Springer Science+Business Media
(www.springer.com)

Supervisor's Foreword

In 2010, the Fermi satellite observed in the GeV energy range two large structures that look like a giant number "8" at the centre of our galaxy, the so-called "Fermi bubbles". The outlines of the bubbles are quite sharp, and the bubbles themselves glow gamma rays uniformly over their colossal surfaces. These regions extend above and below the plane of the Milky Way and these vast apparent shock wave features have been interpreted to be originated by some major disruption from our Galaxy's core. For example, they could have been created by huge jets of accelerated matter blasting out from the supermassive black hole at the centre of our galaxy. Or they could have been formed by a population of giant stars, born from the gas surrounding the black hole, all exploding as supernovae at roughly the same time. In these scenarios injected protons undergo collisions with the ambient matter creating daughter mesons, mostly pions. Therefore, the Fermi bubbles can be expected to be emitters of gamma rays, from the decay of the neutral pions and neutrinos from the decay of charged pions.

For many years, cosmic neutrino detection has been attempted using large volume underwater telescopes. In these detectors the Čerenkov light of the charged lepton produced by neutrino interaction is detected by an array of photomultipliers. The arrival time and charge of the photomultiplier signals are digitised into "hits" and transmitted to shore where an online filter identifies the event. Located in the Mediterranean Sea, the ANTARES experiment is predominantly sensitive to neutrinos from the Southern Hemisphere in the TeV to PeV energy regime. In particular, this allows to study Galactic sources.

In this work, Vladimir Kulikovskiy has analysed the ANTARES data collected in the period 2008–2012 to search for high energy neutrinos in the direction of the Fermi Bubbles. The thesis describes in great detail the different steps followed in his analysis, from the quality of the analysed data sets to the final limit on a possible neutrino emission. The dedicated analysis of the background estimate, which relies on the data themselves, via the study of three off-zone regions, represents an original and effective approach. The procedure of the optimisation of the selection cuts is discussed in detail and allows to set the best limits on the neutrino flux.

The standard recipes to determine the signal significance and the upper limit estimation are reviewed while new calculations based on Bayesian and frequentist's approaches are introduced to include systematic uncertainties and to improve the reliability of the results. A significant part of the work is dedicated to the evaluation of all the possible systematic effects that may arise from inequality of the on- and off-zones and MC simulation ingredients such as water properties and optical module efficiencies. The whole analysis chain can be applied almost without changes to the analysis of the data from extended sky regions measured in other astroparticle detectors.

The present analysis could not establish the emission of high-energy neutrinos from the Fermi bubbles, but leads to various upper limits (depending on the model spectrum considered) which are the most stringent in the world.

Genoa
March 2015

Dr. Marco Anghinolfi

Preface

The ANTARES underwater neutrino telescope was successfully built and now it is operating in the Mediterranean Sea. It has the capability to detect charged particles by Čerenkov light using an array of 875 photomultipliers. Ultra high energetic muon neutrinos interacting with the medium produce muons with almost the same direction and energy as the original neutrinos. The detection of these muons gives information about the astrophysical source that has generated the high energy neutrinos.

The main task of the telescope is the search for cosmic sources of neutrinos. Super Novae Remnants, microquasars and Gamma Ray Bursts are candidates for compact, often point-like neutrino sources, while the Galactic Centre, the Galactic Plane, Fermi bubbles are candidates for extended sources. In particular, the analysis of the Fermi-LAT data has revealed a large structure centred around the Galactic Centre emitting gamma rays—the so-called Fermi bubbles. A promising explanation of this phenomenon is a Galactic wind scenario in which accelerated cosmic rays interact with the interstellar medium in the Fermi bubble regions producing pions. Gamma rays and high-energy neutrinos emission are expected with similar flux from the pion decay.

In this work the ANTARES data during 3.5 years were investigated for a possible emission of the high energetic neutrinos. No statistically significant excess of events was observed. Upper limits on the flux of neutrinos from the Fermi bubbles were derived for various assumed energy cutoffs at the source.

In the first chapter of this manuscript, high energy neutrino astrophysics is introduced. Motivating ideas and evidences of neutrino sources are described in Sect. 1.1. Neutrino detection is described in Sect. 1.2 starting from neutrino interaction with matter and focusing on the optical method of neutrino detection in which the products of interaction are detected by Čerenkov light emitted in the medium. Finally, atmospheric muons and neutrinos are introduced which present the main component of the background noise.

In the second chapter the ANTARES detector is described. Its design overview is done in the first section. The main focus is put on photomuliplier studies since they have been part of this thesis work. Data acquisition and detector calibration are described in the following sections.

In the third chapter the detector simulation is described. The simulation of the detector response to cosmic neutrinos and the atmospheric background is important for physics data analysis. It is used for detector performance estimation, especially, the neutrino direction resolution and energy reconstruction capability. Also, the simulation is used for data quality control and event selection optimisation. A key ingredient of simulations is the sea water optical properties at the ANTARES site which have been described in detail. Another important part of the simulation is photons detection by optical modules. Detailed GEANT simulation of the optical modules was performed and has been part of this thesis work. Finally, environmental optical background is considered at the end of this chapter.

The fourth chapter describes event reconstruction which determines the neutrino direction and energy estimation. The data quality control and the data selection for the main analysis is described at the end of this chapter.

The fifth chapter is devoted to the main analysis of this thesis—a search for neutrino emission from the Fermi bubbles with the ANTARES telescope. The hypothetical neutrino emission is described in the first sections of the chapter. The background estimation method using the off-zones is done in the following ones. Then, the event selection optimisation is presented. Finally, the observed events are presented and compared with the background noise. Upper limits are evaluated using the likelihood elaborated in this thesis for this measurement following a Bayesian approach and compared with the existing methods.

Acknowledgments

I would like to heartily thank Marco Anghinolfi who was my mentor in all aspects from science to way of life and sports for more than 5 years. I had a marvellous time staying in Genova thanks to him and his colleagues. I will miss discussions around the lab's coffee machine, care and facile atmosphere created by Andrea Bersani, Raffaela De Vita, Franco Parodi, Mauro Taiuti, Marco Battaglieri, Mikhail Osipenko, Giacomo Ottonello, Andrea Rottura, Massimo Ivaldi, Ares Ferrari, Katia Fratini, Alessandro Brunengo, Marco Ripani, Paola Montano, Paolo Musico and Marco Brunoldi. All of them helped me a lot by organising and participating in my working process.

Especially I want to thank Heide Costantini. It was exciting to discuss all the new ideas, trying different methods and even redoing a thousand times boring GEANT4 simulations together. Also, I would like to thank Paschal Coyle, Manuela Vecchi, Cristina Di Diorgio, Matteo Sanguineti, Colas Riviere, Jean-Pierre Ernenwein, Rosa Coniglione, Valentina Giordano, Simone Biagi and Maurizio Spurio for support in the Fermi bubbles analysis. It is hard to list all the other people from our ANTARES collaboration who contributed to the analysis by doing the detector construction, maintenance, simulations, data reconstruction, etc. I would like to highlight Veronique Van Elewyck, Laura Core and Katya Lambard for the data quality assessment, Aart Heijboer and Jürgen Brunner for the track reconstruction algorithms, Jutta Schnabel for the events energy reconstruction software, Annarita Margiotta and Colas Riviere for the simulation production, Thomas Eberl for the mass production of data reconstruction and Fabian Schussler for converting both data and simulation to a convenient format for fast analysis. The work of Clancy James and his internal note was very important to understand the water model and the optical photons simulation code. Also, I appreciate the help of Sotiris Loucatos, Juan Jose Hernandez Rey and Simone Biagi for discussions of the statistical approaches. I would like to thank Paschal Coyle, Clancy James, Maurizio Spurio, Piera Sapienza, Veronique Van Elewyck, Antoine Kouchner for careful reading and corrections of the official paper of the analysis. Additionally, I would like to thank some people outside the collaboration: Gary C. Hill, Martin David Haigh and Domenico Costantini for noticing the existing statistical methods and for the help in understanding of these methods as well.

During my Ph.D. I had a chance to work in different labs. I would like to thank Maarten De Jong and Aart Heijboier for the support during my stay at NIKHEF, Antoine Kouchner, Bruny Baret, Alexis Creusot and Veronique Van Elewyck at APC, Vincent Bertin, Heide Costantini, Paschal Coyle, Jürgen Brunner, Colas Riviere, Jean-Pierre Ernenwein, Laura Core and Michel Ageron at CPPM.

My Ph.D. was done in co-doctorate with APC thanks to IDAPP. I would like to thank the IDAPP members for organising summer schools, two-day annual meetings and my stay at APC. Special thanks to Alessandra Tonazzo, Paola Fabbri and Isabella Masina. Thanks to Antoine Kouchner for being my co-tutor and especially for the help in preparing the Ph.D. thesis document. Also, thanks to Benoît Revenu for the thesis manuscript corrections and the discussions at the end of the Ph.D. thesis presentation.

Last year we started to work with FPGAs for KM3NeT, together with Antonio Orzelli and Christophe Hugon. I am really glad to dive into the world of modern electronics with them. Also, the cinema evenings with exchange of our favourite national movies and dishes were amazing.

I will have only good memories from my Ph.D.: A sunset at Institute Michel Pacha, kayaking and having a bath in the sea after the shift days with Marco; climbing in calanques of Marseilles with Colas and Julia; snorkelling and underwater hunting with Heide and Andrea; interesting summer schools where I met a lot of great people and made new friends—Marta and Rémi; all the collaboration meetings, spending the evenings with other ANTARES and KM3NeT members in bars or sightseeing together after the long sessions; skiing with Marco and Matteo after the meetings at CERN; working in Sicily for the KM3NeT, especially doing shift with Nuccio and kite surfing at the beach close to the Capo Passero.

Contents

1 Introduction 1
1.1 High Energy Neutrino Astrophysics 2
1.1.1 Cosmic Rays 2
1.1.2 Possible Sources of High Energy Neutrinos from Gamma Ray Observations 4
1.1.3 Exotic Sources of High Energy Neutrinos 7
1.1.4 Evidence of the Presence of the High-Energy Cosmic Neutrinos 8
1.2 Neutrino Detection 9
1.2.1 Neutrino Interaction 10
1.2.2 Čerenkov Radiation and Track Reconstruction 14
1.2.3 Muons Energy Losses and Energy Reconstruction. . . . 15
1.3 Background Events—Atmospheric Muons and Neutrinos. 18
1.4 Current and Future High Energy Neutrino Detectors 19
References. 21

2 The ANTARES Detector 25
2.1 Detector Design 25
2.1.1 Photomultipliers 27
2.1.2 PMT Choice 29
2.1.3 PMT Characteristics Measurement 31
2.1.4 Optical Module 35
2.1.5 Storey 38
2.1.6 Line Infrastructure 39
2.1.7 Detector Infrastructure 40
2.2 Data Acquisition. 41
2.2.1 Signal Digitisation and Data Transmission 42
2.2.2 Data Filtering and Storage 43
2.3 Detector Calibration 44
2.3.1 Time Calibration. 44
2.3.2 Charge Calibration 45

2.3.3 Position Calibration 46
References 47

3 Simulation of the ANTARES Detector Work 49
3.1 Atmospheric Background Simulation 50
3.2 Neutrino Event Generation 51
3.3 Optical Photons Simulation 54
3.3.1 Optical Properties of Water 54
3.3.2 Implementation of the Optical Photons Simulation . . . 58
3.3.3 OM Simulation 60
3.3.4 Optical Background 62
3.3.5 Optical Background Noise and PMT Electronic Response Simulation 66
References 67

4 Event Reconstruction and Data Selection 71
4.1 Event Reconstruction 71
4.1.1 Track Reconstruction 71
4.1.2 Energy Reconstruction 76
4.2 Celestial Coordinates and Maps 81
4.2.1 Celestial Coordinates 81
4.2.2 Sky Maps 82
4.3 Data Quality 83
4.3.1 Low Level Data Quality 83
4.3.2 Data Quality Using Monte Carlo Simulation 85
References 87

5 A Search for Neutrino Emission from the Fermi Bubbles 89
5.1 Fermi Bubbles 89
5.2 Off-Zones for Background Estimation 91
5.3 Check for the Number of Background Events in the On-Zone and the Off-Zones 92
5.4 Event Selection Criteria 95
5.5 Results 99
5.5.1 Significance Estimation 100
5.5.2 Upper Limits Calculation 104
5.6 KM3NeT 111
References 112

Conclusion 115

Curriculam Vitae 117

Abbreviations

ARS	Analogue Ring Sampler chip (Sect. 2.2.1)
ANN	Artificial Neural Networks (Sect. 4.12)
AVC	Amplitude to Voltage Converter
CPU	Central processing unit (processor)
DAQ	Data AQuisition (Sect. 2.2)
DWDM	Dense Wavelength Division Multiplexing (Sect. 2.1.5)
EMC	Electro Mechanical Cable
FPGA	Field Programmable Gate Array
FWHM	Full Width at Half-Maximum (Sect. 2.1.1)
JB	Junction Box (Sect. 2.1.7)
LED	Light-Emitting Diode
MEOC	Main Electro-Optical Cable (Sect. 2.1.7)
OB	Optical Beacons (Sect. 2.3.1)
OM	Optical Module (Fig. 2.13)
PC	Personal Computer
PCA	Principle Component Analysis
PMT	PhotoMultiplier Tube (Sect. 2.1.1)
RAM	Random-Access Memory
SN	SuperNova
SNR	SuperNova Remnant
SNR MC	SuperNova Remnants with Molecular Clouds
Toy MC	Toy Monte Carlo simulation
TTS	Transit Time Spread (Sect. 2.1.1)

Chapter 1
Introduction

Neutrinos are the lightest known massive particles and they can interact only by weak interaction. These properties allow them to act as a unique transmitter of information. They can pass through the interstellar matter, escape from the dense cores of astrophysical sources and travel without any deflections even through strong magnetic fields. Neutrinos provide information complementary to photons and astronomers aim to use these probes to look deeper in space, look behind the sources or inside them. However the detection of neutrinos is much more difficult with respect to hadrons and photons and gigaton masses detector are needed to detect astronomical sources with their expected neutrino fluxes.

In addition to the possibility to find new sources, neutrino telescopes will allow to understand the production mechanism of cosmic rays in space. There are two competing hypothesis for most of the discovered gamma ray sources: leptonic and hadronic models. In leptonic models, mostly electrons are accelerated while in hadronic models protons and heavier ions are accelerated. Gamma rays can be produced in both models: by the electrons via inverse Compton process or by the decaying pions from accelerated protons interacting with matter surrounding the source. Pions are expected to produce a similar quantity of neutrinos and photons. On the contrary, no neutrinos are expected from electrons interaction. Neutrino detection from an astrophysical source will be a smoking gun for hadronic models. There are several gamma ray sources above TeV which make the neutrino astronomy promising.

Neutrinos produce a charged lepton (electron, muon, tau) and an hadronic shower through weak interaction with matter. Registration of these secondary particles is possible thanks to the Čerenkov light. This light is emitted during the passage of charged particles with velocities higher than speed of light in the medium. Among the charged particles the muons have the longest free path in the matter, roughly 4 m per 1 GeV in the water which allows a high precision track direction reconstruction. The muon track deviates from the neutrino one with an angle less than 1° at energies above 2 TeV. These features make high energy muon neutrinos ideal particles to be used in neutrino telescopes. In addition to the Čerenkov light emission alternative detection methods are used. In particular, at ultra-high energies (above PeV) neutrinos can be detected through the acoustic and the radio waves emission from the hadronic shower produced in the interaction.

V. Kulikovskiy, *Neutrino Astrophysics with the ANTARES Telescope*,
Springer Theses, DOI 10.1007/978-3-319-20412-3_1

There is a working neutrino telescope in the ice at the South Pole (IceCube Ahrens et al. 2001) of roughly 1 km^3 size and there are two smaller working detectors in the water: the ANTARES detector (Ageron et al. 2011) in the Mediterranean sea, which is the object of this thesis, and the BAIKAL detector (Belolaptikov et al. 1997) in the lake Baikal in Russia. Finally, the KM3NeT collaboration aims to build a cubic kilometres size neutrino telescope in the Mediterranean sea (Aiello et al. 2011).

1.1 High Energy Neutrino Astrophysics

Study of the low energy neutrino emission from SN1987A explosion has become a classical part of the university books (Bahcall 1989, ch. 15.5). There are several strong evidences that the high energy neutrino astrophysics is possible too: cosmic rays of $E \sim 10^{20}$ eV and different sources of high-energy gamma rays were observed. The presence of the high energy cosmic rays is important since their interactions with the interstellar medium or molecular clouds produce mesons. Those in turn decay with production of gamma rays and neutrinos. In particular, the following reactions take the biggest role:

$$\pi^0 \rightarrow \gamma + \gamma \tag{1.1}$$

$$\pi^+ \rightarrow \mu^+ + \nu_\mu \tag{1.2}$$

$$\pi^- \rightarrow \mu^- + \tilde{\nu}_\mu \tag{1.3}$$

These reactions show that the production of the neutrinos happens together with the gamma rays. The opposite, however, is not true as gamma rays may appear from a bremsstrahlung or an inverse Compton scattering processes valid for relativistic electrons. These brings to the phenomenological distinction between the models: in the hadronic acceleration model gamma rays and neutrinos are produced together, in the leptonic acceleration model, instead, only gamma rays are produced.

1.1.1 Cosmic Rays

The cosmic-ray spectrum shown in Fig. 1.1, measured with air shower detectors, demonstrates that protons and heavier nucleons with energies in the very broad range extending to 10^{20} eV arrive to the Earth. Only charged particles can be accelerated, in particular, protons and electrons. If protons or nuclei in general are accelerated in a source, then due to the interaction with the matter surrounding the source cosmic rays should be produced together with neutrinos. The presence of the cosmic rays of high energies, together with the assumption that their primary particles were hadrons, makes high energy neutrino detection theoretically possible.

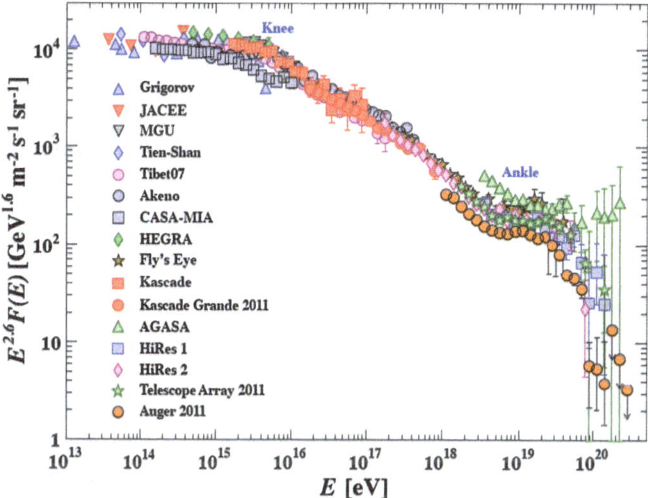

Fig. 1.1 Cosmic-ray spectrum as a function of E (energy-per-nucleus) from air shower measurements (Beringer et al. 2012, ch. 26.5). Two changes in spectrum are clearly seen (knee and ankle)

The question where the high energy cosmic rays are accelerated still remains one of the interesting unresolved problem in the astroparticle physics. Energy spectrum of the cosmic rays is decreasing with a power law with two changes on the power spectrum coefficient. At several PeV it becomes steeper. This point is called knee (Fig. 1.1). It is supposed that sources in our Galaxy are able to accelerate particles till the energies around the knee so the cumulative distribution has this steepening till the so-called ankle (about 3×10^{18} eV). At the energies above the ankle Larmor radius of the protons in the Galactic magnetic field is larger than the Galaxy radius so the protons cannot be confined in the Galaxy. The flattening of the spectrum in this region may be associated with the onset of the extra-Galactic component. An alternative explanation is that the transition from galactic to extragalactic cosmic rays happens at a much lower energy and the ankle in this model is produced by the modification of the extragalactic source spectrum of primary protons interacting with the cosmic microwave background (Aloisio et al. 2008). At energies above 4×10^{19} eV a suppression of the flux with respect to a power law extrapolation is found, which is compatible with the predicted Greisen–Zatsepin–Kuzmin effect, but could also be related to the maximum energy that can be reached at the sources (Abraham et al. 2010).

To study the source accelerating the cosmic rays, it could be extremely efficient to localise it on the sky by detecting the high energy particles and compare the observations with the optical telescopes. For the air-shower detectors this is an hard task due to the fact that the primary charged particles themselves are deflected in the magnetic fields (Earth's and interstellar). The pointing accuracy is still possible for such instruments at extremely high energies (about 10^{18} eV) or based on neutron detection. The detection of a cosmic-ray point source would directly open a possibility for the neutrino detection from the same source. However, the recent analysis of the Pierre Auger surface detector has not revealed such sources yet (Abreu et al. 2012). For the gamma-ray detectors the pointing accuracy is achievable at much lower energies. Several very interesting compact sources are recently revealed with gamma-ray detectors and their impact on the neutrino astronomy is described in the next section.

1.1.2 Possible Sources of High Energy Neutrinos from Gamma Ray Observations

A most of the sources detected by the gamma-ray detectors are the supernova remnants surrounded by the molecular clouds (SNR MC). The measurements of the gamma-ray spectrum with the Fermi-LAT satellite (Ackermann et al. 2013) are shown in Fig. 1.2 for the two SNRs: W44 and IC 443. The measurements in the sub-GeV range allow to distinguish the origin of the gamma rays in these sources. The fit with the gamma-ray spectrum originating from the π^0 decay describes the data well in contrast to the bremsstrahlung gamma-ray spectrum induced by electrons. This makes the hadronic models preferable in respect to the leptonic ones.

The idea of the explaining cosmic rays below the knee with the shock acceleration mechanisms in SNRs becomes very popular in the last years. However, energy of the gamma rays detected in the SNRs is limited till tens of TeV. Also recent theoretical calculations and Monte Carlo simulations still can not reach PeV thresholds of the cosmic-ray energies.

High-energy gamma-ray sources are promising neutrino sources as well. These sources include the mentioned W44 and IC 443, other SNR MC as RX J1713.7-3946, Vela Jr, Vela X, together with the several other types of the sources such as Cygnus Region which is a star forming region. The ideal case would be that only hadrons are accelerated in these objects (pure hadronic model). In this case all the gamma rays are coming together with a similar amount of neutrinos. Their exact quantity may be calculated from the measured gamma-ray flux. Calculations show that size of the detectors should be of the order of kilometres (Vissani and Aharonian 2012).

The sources with an apparent size on the sky smaller than the detector angular resolution are indistinguishable from the point-like. This can be applied to the above mentioned sources which seem point-like or slightly extended for the detectors with angular resolution of about one degree. In contrast to them, there are several known largely extended source candidates. The centre of our Galaxy is an area of the active star formation. In this region supernovae (SN) explosion rate is very high which makes this region interesting for neutrino detection. Another possible extended source of the neutrinos is the so-called Fermi bubbles region which is shown in Fig. 1.3. The study of the neutrino emission from this region was the main topic for my thesis and it will be described in details in Sect. 5.1. Both Galactic Centre and Fermi bubble regions have a significant high-energy gamma-ray emission.

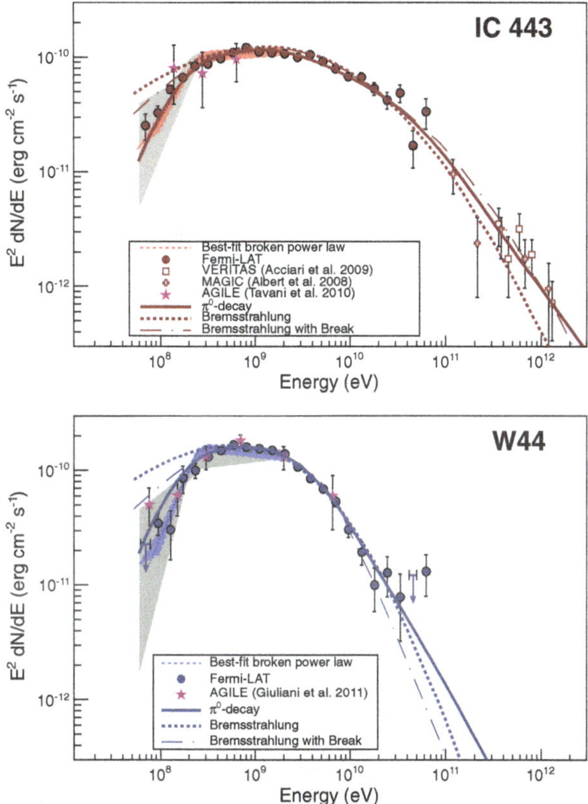

Fig. 1.2 Gamma ray spectra of IC 443 (*top*) and W44 (*bottom*) as measured with the Fermi-LAT. Color-shaded areas bound by the *dashed lines* denote the best-fit broadband smooth broken power law (from 60 MeV to 2 GeV), *gray-shaded bands* show systematic errors below 2 GeV due mainly to imperfect modelling of the galactic diffuse emission. *Solid lines* denote the best-fit pion-decay gamma-ray spectra, *dashed lines* denote the best-fit bremsstrahlung spectra, and *dash-dotted lines* denote the best-fit bremsstrahlung spectra when including an ad hoc low-energy break at 300 MeV in the electron spectrum. The measurements from other gamma ray detectors are shown also. The figure is taken from (Ackermann et al. 2013)

Another promising sources are the Gamma Ray Bursts (GRBs): the transient flashes of the gamma rays which have a duration of the range from sub-seconds up to several hundred seconds. They were first discovered in 1960s by the U.S. Vela satellites, which were built to detect the gamma-ray pulses emitted by the nuclear weapons tested in the space. With the detection of the over 8000 GRBs, by using the BATSE experiment it became clear that the GRB distribution on the sky is isotropic. That suggests an extragalactic origin of the GRBs. Several models predict a burst of high-energy neutrinos in a concurrence with the flash of the gamma rays, also referred

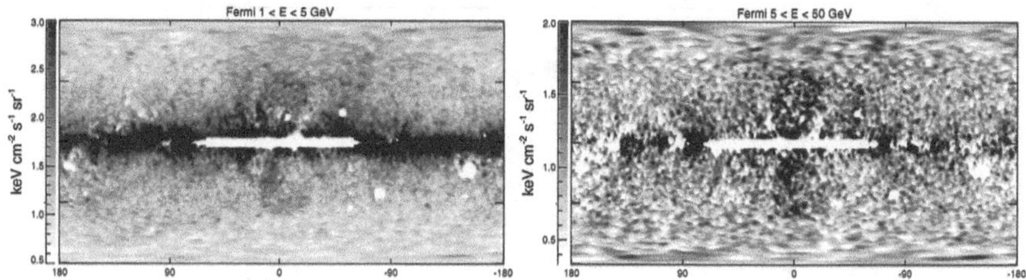

Fig. 1.3 Full sky residual maps after subtracting the dust and galactic disk templates from the Fermi-LAT 1.6 year gamma-ray maps in two energy bins. Point sources are subtracted, and large sources, including the inner disk ($-2° < b < 2°$, $-60° < l < 60°$), have been masked. Two large bubbles are seen spanning ($-50° < b < 50°$) in both cases. The image is taken from (Su et al. 2010)

to as a prompt emission (Mészáros 2006). In these models, neutrinos are produced by interactions of the accelerated by the shock waves protons with energetic ambient photons.

Neutrinos are produced in hadronic models by the pion decays (1.3) and, in turn, by the muon decays:

$$\mu^- \rightarrow e^- + \bar{\nu}_e + \nu_\mu \tag{1.4}$$

$$\mu^+ \rightarrow e^+ + \nu_e + \bar{\nu}_\mu. \tag{1.5}$$

According to this simplified chain, neutrinos of the three flavours produced with ratio $1:2:0$ ($\nu_e : \nu_\mu : \nu_\tau$). Neutrino oscillations, however, change the original composition on a way to observer. For the transparent sources (excluding oscillations in matter) the oscillation probabilities are the following:

$$P_{l\rightarrow l'} = \left| \sum_{i=1}^{3} U_{li}^* U_{l'i} e^{-im_i^2 L/2E} \right|^2, \tag{1.6}$$

where U_{li} are the elements of the neutrino mixing matrix. Expansion of this expression gives sinus terms with $\Delta m_{ij}^2 L/2E$ phases. One can introduce the oscillation length:

$$L = \frac{4\pi E}{\Delta m_{ij}^2} \tag{1.7}$$

For the sources much larger than the oscillation length or for the groups of the sources parsed over the large range of the distances, the oscillation probabilities are averaged and may be expressed as (Bilenky and Pontecorvo 1978):

$$P_{ll'} = \sum_{i=1}^{3} |U_{li}^2||U_{l'i}^2| \quad l, l' = e, \mu, \tau. \tag{1.8}$$

Using recent measurements of the neutrino oscillations parameters, mixing probability matrix $P_{ll'}$ may be estimated as (Vissani and Aharonian 2012):

$$P_{l,l'} = \begin{pmatrix} 0.543 & 0.262 & 0.195 \\ 0.262 & 0.364 & 0.374 \\ 0.195 & 0.374 & 0.430 \end{pmatrix} \qquad l, l' = e, \mu, \tau, \tag{1.9}$$

with uncertainties of ∼10 % on each value. With this matrix oscillated flux becomes:

$$\begin{pmatrix} 1 & 2 & 0 \end{pmatrix} \times \begin{pmatrix} 0.543 & 0.262 & 0.195 \\ 0.262 & 0.364 & 0.374 \\ 0.195 & 0.374 & 0.430 \end{pmatrix} = \begin{pmatrix} 1.067 & 0.990 & 0.943 \end{pmatrix}, \tag{1.10}$$

which means that the arriving neutrinos ν_e, ν_μ, ν_τ are expected to come approximately with the same fluxes. For GeV neutrinos the oscillation length (1.7) is of the order of hundreds of kilometres or 10^{-10} parsec which makes the proportion (1.10) applicable for the majority of the sources.

1.1.3 Exotic Sources of High Energy Neutrinos

The existence of the Dark Matter is well established (Beringer et al. 2012, ch. 24). The first evidence, which is also the most convincing, comes from velocities of stars and other luminous objects. For some of these objects the measured velocities are too high in comparison to the visible mass around which the objects are rotating. This suggests a presence of the invisible mass. In particular, the so-called dark halo in the centre of our Galaxy with a mass density $\rho \sim 1/r^2$, fits well the observations.

One of the other interesting evidences is the observation of the bullet galaxy cluster 1E0657-558 (Tucker et al. 1995) which passed through another galaxy cluster. The X-ray data analysis shows that the visible baryonic mass of the gas was decelerated while the galaxies in the cluster continued to move following the dark matter. This fact also hints the hypothesis that the dark matter is weakly self-interacting as it would annihilate or decelerate in the opposite case (Markevitch et al. 2004).

The measurements of the total amount of the dark matter in the Universe comes from the global fit of several different types of independent measurements: the Cosmic Microwave Background, the Hubble constant measurements with the "standard candles"—supernovae type 1a and the Baryonic Acoustic Oscillations (Beringer et al. 2012, Ch. 24). The confidence area for cosmological constant Ω_Λ and total mass Ω_m from these measurements is shown in Fig. 1.4. First, it is interesting to see that all three confidence area are crossing in the same zone, which is also consistent with the flat Universe hypothesis ($\Omega_\Lambda + \Omega_m = 1$). The total mass Ω_m consists of the baryonic mass Ω_b which is now measured with quite high accuracy from

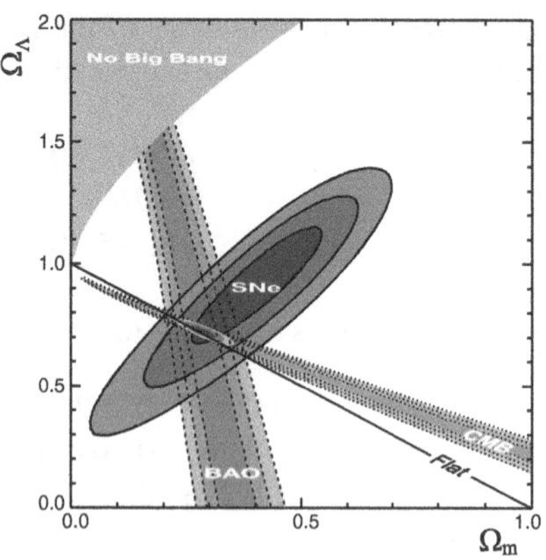

Fig. 1.4 Confidence level contours of 68.3, 95.4 and 99.7 % in the $\Omega_\Lambda \Omega_m$ plane from the CMB, BAOs and the Union SNe Ia set, as well as their combination. The image is taken from Kowalski et al. (2008)

the CMB and large-scale structure, and is consistent with the determination from Big-Bang nucleosynthesis predictions; the range is $0.019 \leq \Omega_b h^2 \leq 0.024$ (95 % confidence level), where h is the Hubble parameter, $h = 0.704 \pm 0.025$ (Beringer et al. 2012, Ch. 22). Total expected amount of light neutrinos is expected to have a negligible contribution to Ω_m ($\Omega_\nu h^2 \leq 0.0062$ 95 % confidence level (Beringer et al. 2012, Ch. 24). The rest (>70 %) is ascribed to the dark matter.

Simulations of the matter in the Universe shows that the dark matter can not be relativistic in order to create the cosmic structures similar by the shape and the size to the observed ones (galaxies, clusters, filaments). Dark matter could be easily accumulated in the massive objects such as the Sun and the Earth. Decay or annihilation of the dark matter could produce neutrinos. So, even the astronomical objects which does not accelerate particles could be the sources of high energy neutrinos.

1.1.4 Evidence of the Presence of the High-Energy Cosmic Neutrinos

Recently, two events with the energies exceeding 1 PeV were discovered by the IceCube collaboration (Aartsen et al. 2013b). These events have a topology of the cascades initiated by the neutrinos inside the detector volume. The probability of observing two or more candidate events of such energies under the atmospheric background-only hypothesis is 2.9×10^{-3} which corresponds to 2.8 σ of the discovery of cosmogenic origin of this two neutrino events. Lately, this search was extended to the lower energies and 26 events were observed additionally to the already seen two. The energy distribution of the events together with expectation from the atmospheric background is shown in Fig. 1.5. These events lead to 4.1 σ significance of the presence of the cosmic neutrinos. The best-fit energy spectrum

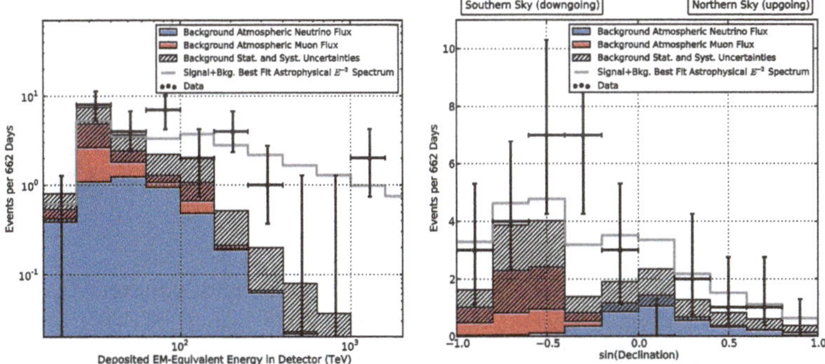

Fig. 1.5 Distribution of the deposited energies (*left*) and declination angles (*right*) of the observed events compared to model predictions. Energies plotted are in-detector visible energies, which are lower limits on the neutrino energy. Note that deposited energy spectra are always harder than the incident neutrinos spectrum. This is effect is due the neutrino cross-section increasing with energy. The expected rate of atmospheric neutrinos is based on northern hemisphere muon neutrino observations. The estimated distribution of the background from atmospheric muons is shown in *red*. Due to lack of statistics from data far above the cut threshold, the shape of the distributions from muons in this figure has been determined using Monte Carlo simulations with total rate normalised to the estimate obtained from an in-data control sample. Combined statistical and systematic uncertainties on the sum of backgrounds are indicated with a hatched area. The *grey line* shows the best-fit E^{-2} astrophysical spectrum with all-flavour normalisation (1:1:1) of $E^2\Phi_\nu(E) = 3.6 \times 10^{-8}$ GeV cm^{-2} s^{-1} sr^{-1} and a spectral cutoff of 2 PeV. The image is taken from Aartsen et al. (2013a)

of the extraterrestrial component is similar ($E^{-2.2}$) to the E^{-2} spectrum generically expected from a primary cosmic ray accelerator. For such a generic E^{-2} spectrum a best-fit is obtained with $E^2\Phi(E) = (3.6 \pm 1.2) \cdot 10^{-8}\,\mathrm{GeV\,cm^{-2}\,s^{-1}\,sr^{-1}}$ total neutrino flux normalisation and cutoff energy at PeV energies (Aartsen et al. 2013a).

1.2 Neutrino Detection

Neutrinos are weakly interacting particles. Their detection is possible by an identification of the interaction products with the matter. As the cross-sections of the interactions are on the weak scale (Sect. 1.2.1), large detector volumes are required. The interactions can be detected using the light, radio or acoustic emission.

- The light is emitted mostly due to the charged products passage through the matter (Čerenkov radiation) which is used in the current work and it is described in Sect. 1.2.2; description of the light detection using photomultipliers is done in Sect. 2.1.1.
- The radio emission due to the Askaryan effect is similar to the Čerenkov radiation: a particle traveling faster than the phase velocity of light in a dielectric (such as salt, ice, air or the lunar regolith) produces a shower of secondary charged particles

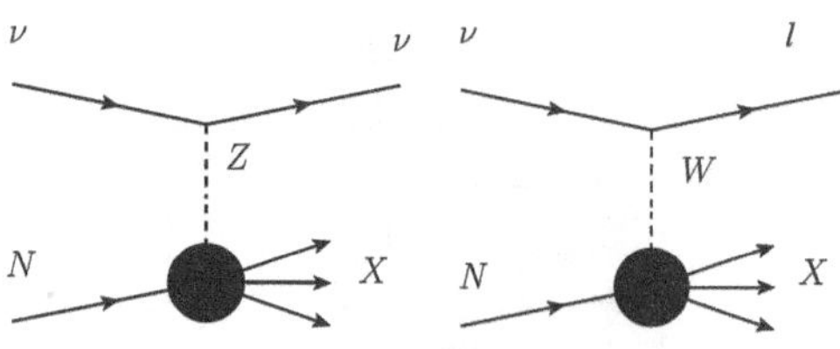

Fig. 1.6 Neutral current interactions (*left*) and charged current interactions (*right*)

which contain a charge anisotropy and thus emits a cone of coherent radiation in the radio or microwave part of the electromagnetic spectrum. This effect was first predicted in 1962 and lately it was observed in several materials (including ice Gorham et al. 2007 and air Marin et al. 2011). Nowadays several experiments are in construction or started collecting the data at South Pole.

- Acoustic emission happens due to the instantly heated matter at the local area of the interaction and the fast expansion with a propagation of the pressure wave. The acoustic registration is promising in the water and several steps are done towards a detector creation (Aguilar et al. 2011).

1.2.1 Neutrino Interaction

Neutrinos participate in interactions by exchange of weak bosons. Interactions with an exchange of $W^{\pm}$ boson are called charged current (CC) opposite to the neutral current (NC) interactions via Z boson. Both reactions are presented schematically in Fig. 1.6. In the first reaction a high energy lepton is produced. Variety of neutrino interaction processes is described by the products of interaction with a nucleus. At the low neutrino energies it can be a nucleus excitation, an atomic number change or a fission. At higher energies a different hadrons production is possible.

In the range $E_\nu \sim 0 - 1\,\text{MeV}$ there are two thresholdless processes:

Coherent scattering It involves the neutral-current exchange where a neutrino interacts coherently with the nucleus:

$$\nu + {}^{A}_{Z}X \rightarrow \nu + {}^{A}_{Z}X^{*}, \tag{1.11}$$

where A is mass number (number of nucleons) and Z is atomic number (number of protons).

Neutrino capture on radioactive nuclei Sometimes it is referred to as enhanced or stimulated beta decay emission. The process is similar to that of ordinary beta decay with a difference that the neutrino is interacting with the target nucleus:

$$\nu_e + {}^{A}_{Z}X \rightarrow e^{-} + {}_{Z+1}^{A}X. \tag{1.12}$$

Whereas coherent scattering happens on nucleus as a single coherent structure, energies $E_\nu \sim 1 - 100\,\text{MeV}$ allows access to nucleons individually. These reactions depend on the targets and can be summarised as following:

Antineutrino-proton scattering or inverse beta decay:

$$\bar{\nu}_e + p \rightarrow e^+ + n. \tag{1.13}$$

This reaction allowed to confirm the neutrino existence (Reines and Cowan 1997). The products of the reaction can be well detected as a two light flashes arriving in coincidence: one is due to the positron annihilation (together with an electron quickly found in the matter it produces two gamma rays) and a slightly delayed signal from the photon emission due to the neutron capture by cadmium for example. Also, the reaction (1.13) plays an important role in understanding supernova explosion mechanisms.

Neutrino-deuterium interactions:

$$\nu_e + d \xrightarrow{\text{CC}} e^- + p + p, \tag{1.14}$$

$$\nu + d \xrightarrow{\text{NC}} \nu + n + p. \tag{1.15}$$

The first one goes via charged current and it is available only for electron neutrinos at $E_\nu < m_\mu$. The second, instead, goes with neutral current and it is the same for all neutrino flavours. Both reactions allowed the SNO experiment with a heavy water target to simultaneously measure the electron and non electron component of the solar neutrino spectrum and confirm the neutrino oscillations (Ahmad et al. 2001).

Other nuclear targets The reactions are basically the following:

$$\nu(\tilde{\nu}) + {}^{\text{A}}_{\text{Z}}\text{X} \rightarrow l^{\mp} + {}_{\text{Z}\pm 1}{}^{\text{A}}\text{X}, \tag{1.16}$$

where lepton l corresponds to the neutrino flavour. This reaction is of particular importance for the first solar neutrino detection using a ^{37}Cl target (Cleveland et al. 1998).

As the neutrino energy grows the description of a neutrino scattering becomes more diverse. In the intermediate energy range $E_\nu \sim 0.1 - 20\,\text{GeV}$ there are three main categories:

Elastic and quasielastic scattering Neutrinos can elastically scatter off an entire nucleon liberating a nucleon (or multiple nucleons) from a target. In the case of charged current neutrino scattering, this process is referred to as "quasielastic scattering" whereas for neutral-current scattering this is traditionally referred to as "elastic scattering".

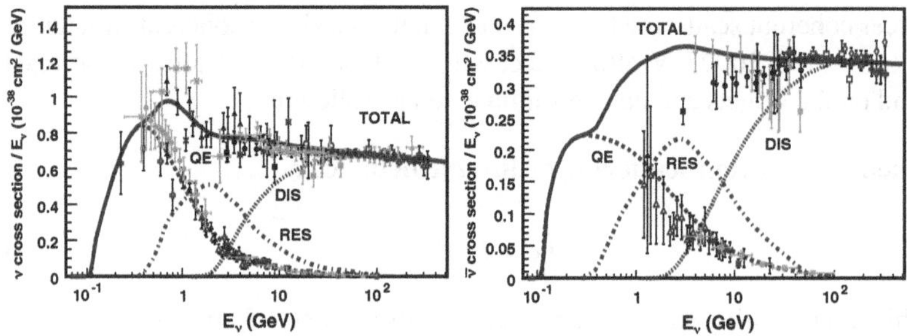

Fig. 1.7 Total neutrino and antineutrino per nucleon CC cross-sections (for an isoscalar target) divided by neutrino energy and plotted as a function of energy. Points with markers present different measurements. Predictions for different contributions are shown: quasielastic scattering (*dashed*), resonance production (*dot-dashed*), and deep inelastic scattering (*dotted*). This figure is taken from Formaggio and Zeller (2012)

Resonance production Neutrinos can excite the target nucleon to a resonance state. The resultant baryonic resonance (Δ, N^*) decays to a variety of possible mesonic final states producing combinations of nucleons and mesons.

Deep inelastic scattering Given enough energy, the neutrino can resolve the individual quark constituents of the nucleon. This is called deep inelastic scattering and manifests in the creation of a hadronic shower.

The total cross-sections for these processes are shown in Fig. 1.7. At high energies the inclusive cross-sections of the deep inelastic scattering on hadrons dominates over the other possible interactions. Neutrinos with $E_\nu > 10\,\text{GeV}$ interact with the quarks and gluons inside nucleons. These components of the nucleons were originally called partons. The cross-section can be estimated by the following equation (Beringer et al. 2012, (44.18, 44.19)):

$$\frac{\mathrm{d}^2\sigma}{\mathrm{d}x\mathrm{d}y}(\nu X \to l^- Y) = \frac{G_F^2 xs}{\pi}[(d(x) + ...) + (1-y)^2(\bar{u}(x) + ...)] \quad (1.17)$$

and for antineutrinos:

$$\frac{\mathrm{d}^2\sigma}{\mathrm{d}x\mathrm{d}y}(\bar{\nu} X \to l^+ Y) = \frac{G_F^2 xs}{\pi}[(\bar{d}(x) + ...) + (1-y)^2(u(x) + ...)], \quad (1.18)$$

where $x = Q^2/(2M\nu)$ is a fraction of the nucleon's energy carrying by the quark, $\nu = E - E'$ is the energy lost by the lepton in the nucleon rest frame, $1 - y = E'/E$, $s = 2ME$ is the center-of-mass energy squared of the lepton-nucleon system and $Q^2 = 2(EE' - \vec{k}\vec{k}') - m_\nu - m_l$ ($\vec{k}$ and $\vec{k}'$ are the initial and the final momentum of the neutrino). The quark distribution functions $u(x)$, $\bar{u}(x)$, $d(x)$, $\bar{d}(x)$ represent densities of the partons carrying momentum fraction x inside the nucleon. The contribution of the higher mass quarks was suppressed in these equations. At low Q^2

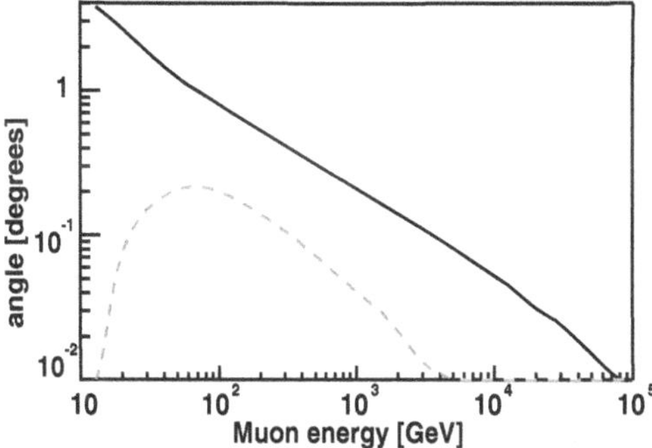

Fig. 1.8 Angle median distribution of the produced muons arriving to the detector and the neutrinos with E^{-2} spectrum (*solid line*). Distribution of the median of the angle between the same muons at the interaction point and arriving to the detector (*dashed line*). At low energies most of the muons start within the detector volume, so the effect of muon scattering in the medium is low. The image is taken from Bailey (2002)

parton distribution functions are dominated by the tree valence quarks (*uud* for *p* and *udd* for *n*). At high Q^2 there are more quark-antiquark pairs,—the so-called sea quarks. Also, at the high energies quarks and antiquarks carry only about a half of the nucleon momentum, the rest is carried by the gluons. By using the Quantum Chromodynamics aimed to describe the strong interactions between quarks and gluons, it is difficult to obtain the parton distribution functions with a desired precision because the factors entering the calculations become non-perturbative. Parton densities functions, however, are universal between different reactions. This opens a possibility to measure them in accelerator experiments, for example, fixed target lepton-nucleon scattering experiments at SLAC, FNAL, CERN and electron-proton collider HERA at DESY.

As it can be seen from (1.17), (1.18), the cross-section grows rapidly with a decrease of $E - E'$ and increase of E'/E. So the cross-section is much higher when the charged lepton brings out most of the incident neutrino energy. Including the momentum conservation law this means that in most cases the produced lepton has an energy and a direction similar the neutrino's ones. Median of the angle between the muon and the neutrino evaluated with the simulation described is present in Fig. 1.8.

Muons, due to their larger mass in respect to the electron and longer life time in respect to the tau, have the longest tracks in the medium. Muons with TeV energies can pass through kilometres of water (Fig. 1.9). Their passage causes a Čerenkov light, e^+e^- pair production, a bremsstrahlung radiation and photonuclear interactions. If the emitted light can be collected in a way to locate its origin, the track can be reconstructed.

Hadronic products of the interaction form an hadronic shower. Size of the shower is rather little in respect to the muon track which makes it difficult to detect. High energy electrons instead of muons have a very high probability to radiate bremsstrahlung photons after a few centimetres of water. This immediately creates an electromagnetic

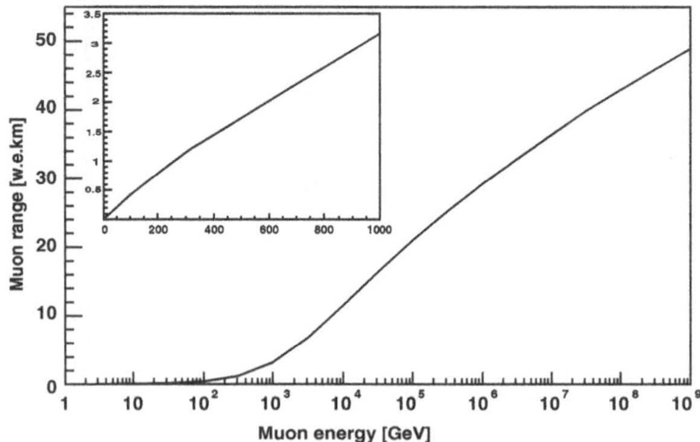

Fig. 1.9 Muon maximum ranges in the water obtained from simulation. The image is taken from Bailey (2002)

shower due to the $\gamma \rightarrow e^+e^-$ pair production. For a 10 TeV electron, the shower length is only about 10 m. For the tau leptons due to their short life time it is possible, in principle, to see the double shower structure with the first shower coming from the interaction point of the incident neutrino and the second one due to the tau decay. These two showers should be separated by the tau decay length $l_\tau = \gamma c t_\tau \approx 50(E_\tau/\text{PeV})$ m.

Also it is possible in principle to detect a sound which is created by the expanding shock wave of the heated medium due to the neutrino interaction. This method seems to be very promising for the energies above 10^{20} eV.

So far for the energy range of 100 GeV–1 EeV the detection of muons by the Čerenkov radiation is the most precise and efficient method for the neutrino telescopes.

1.2.2 Čerenkov Radiation and Track Reconstruction

When the charged particle passes through the matter it polarises its atoms creating electric dipoles. If the particle moves with a speed slower than the speed of light in the medium, these dipoles are symmetrically oriented around the track. At higher particle velocities this symmetry is broken and the Čerenkov light is emitted due to the dipole radiation. The light is emitted at a fixed angle in respect to the particle momentum.

A charged particle traveling with velocity $u > c/n$ emits spherical waves of light along its trajectory. The spherical waves emitted at every point on the trajectory create collectively a wavefront with a cone shape. Figure 1.10 can be used to describe the Čerenkov angle: the spherical wave emitted at the point A will reach the point C at the same time as the charged particle arriving time at the point D. The cosine of the angle θ_c is given by:

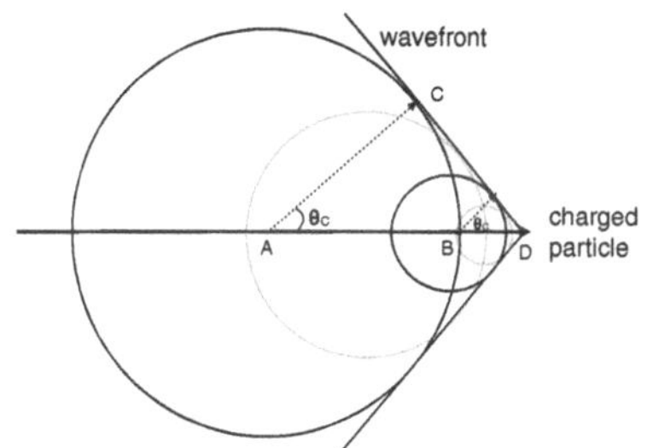

Fig. 1.10 Scheme of the Čerenkov radiation. Emission of the Čerenkov light from each point along the track has a spherical wavefront which produces a *cone-shape* wavefront for the whole track

$$\cos\theta_c = \frac{AC}{AD} = \frac{tc/n}{t\beta c} = \frac{1}{\beta n}. \tag{1.19}$$

Assuming for relativistic particles ($\beta \approx 1$) and $n = 1.33$ for the water the value of this angle is about 41.2°. The number of Čerenkov photons emitted per track length x and wavelength λ can be estimated using the following equation:

$$\frac{d^2N}{dxd\lambda} = \frac{2\pi\alpha Z^2}{\lambda^2}(1 - \frac{1}{\beta^2 n^2}), \tag{1.20}$$

where Z is the charge of the particle and α is the electromagnetic coupling constant.

For the single photon detection with a good timing resolution it is efficient to use photomultiplier tubes (PMTs). Their principle of work is described in detail in Sect. 2.1.1. In the sensitive diapason of the photomultiplier tubes 350 nm $\leq \lambda \leq$ 600 nm which are used in the ANTARES detector almost 200 emitted photons per centimetre of track can be expected according to (1.20).

If the produced Čerenkov light is seen in several places of the detector medium then the track reconstruction is possible by analysing a time and coordinates of the arrived photons. Densely placed detection units allow more photons to be detected and the better track reconstruction correspondingly. The example[1] of the track reconstruction in the ANTARES detector is shown in Fig. 1.11.

1.2.3 Muons Energy Losses and Energy Reconstruction

The processes by which muons loose energy as they propagate through matter can be summarised as following:

Ionisation The charge interactions occur in single collisions, leading to ionisation, atomic, or collective excitation. For 90 % of all collisions the energy losses are less than 100 eV (Beringer et al. 2012), which creates smooth and continuous

[1] http://www.pi1.physik.uni-erlangen.de/antares/online-display/online-display.php.

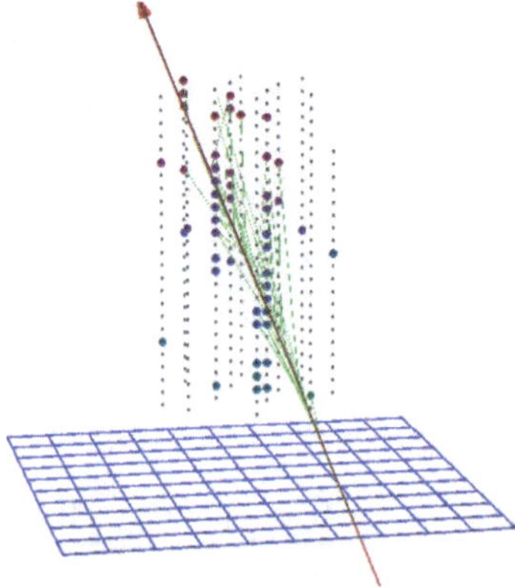

Fig. 1.11 Muon track (*red arrow*) reconstruction in the ANTARES detector. ANTARES light detection units (optical modules) are shown as *grey points*. *Colour points* represent the optical modules which detected photons. Emission of the Čerenkov light is extrapolated with the *green lines* for such modules. The sea bed is schematically shown as a *blue grid*. The event is taken from the real data online display

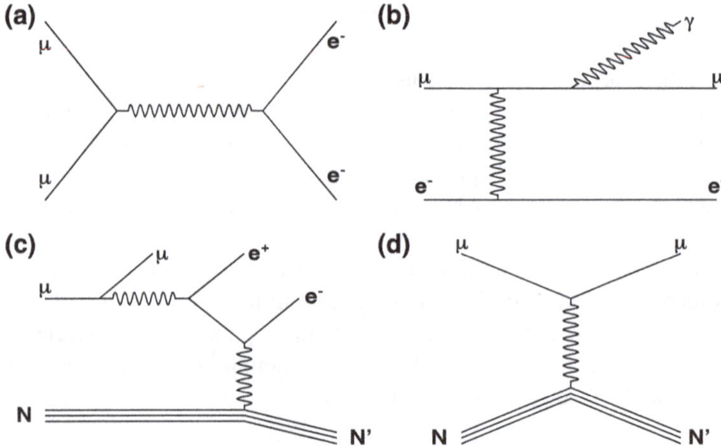

Fig. 1.12 Muon energy loss processes in matter: muon scattering from atomic electrons (**a**), bremsstrahlung on atomic electrons (**b**), electron pair production (**c**) and photonuclear interaction with a general nucleon (**d**). The image is taken from Bailey (2002)

energy transfer. These losses can be calculated using the Bethe-Bloch formula and the basic process is shown in Fig. 1.12a.

Bremsstrahlung In the electric field of a nucleus or atomic electron, muons can radiate high energy photons. This process is shown in Fig. 1.12b.

Pair production A muon can radiate a virtual photon which, again in the electric field, can convert into e^+e^- pair. This process is shown in Fig. 1.12c.

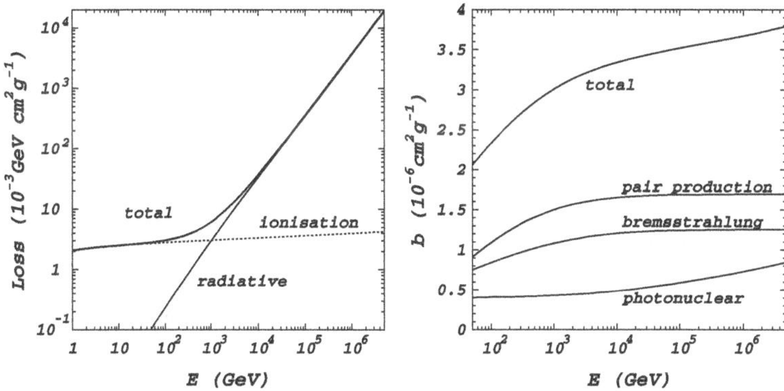

Fig. 1.13 Muon energy loss in pure water as a function of its energy. Transition of the losses due to ionisation to the losses due to radiative processes (*left*). Contributions to the radiative energy losses in the pure water: e^+e^- pair production, bremsstrahlung, and photonuclear interactions (*right*). The image is taken from Klimushin et al. (2001)

Photonuclear interactions A virtual photon from muon directly interacts with a nucleus. The interaction is either pure electromagnetic or it becomes more complicated if virtual photon converts to a ρ^0 virtual vector meson. The basic diagram is shown in Fig. 1.12d.

Knock on electrons The diagram is the same as for the ionisation losses, however the energy gained by electron is enough to leave the atom and even produce Čerenkov light while traveling through the matter.

Ionisation energy losses are dominant at low energies and they are fairly constant and homogeneous over the track. The value is about 2 MeV/cm at 1 GeV in water. At energies above 100 GeV the main energy losses are due to the radiative processes like e^+e^- pair production, bremsstrahlung and photonuclear processes. These losses are strongly energy dependent and stochastic. Therefore, only average total energy loss can be calculated. The value and composition of energy losses for different muon energies is shown in Fig. 1.13. It is convenient to express the average energy loss as:

$$-\left\langle \frac{dE}{dX} \right\rangle = a(E) + b(E)E. \tag{1.21}$$

This equation emphasise the fact that, unlike the ionisation losses $a(E)$, the radiative energy losses $b(E)E$ are proportionally increasing with energy. It is worth to mention that radiative processes produce hadronic and electromagnetic cascades along the muon track.

The following properties of the energy losses make possible muon energy reconstruction: for the low muon energies the track length is proportional to the particle energy, for the high energies the amount of the detected light per track length may be used instead.

1.3 Background Events—Atmospheric Muons and Neutrinos

Cosmic rays interacting with the atmosphere produce mesons and baryons which may decay with a production of leptons. The leptons direction and energy is different from the original cosmic ray ones. Muons and neutrinos among these leptons constitute the background noise for a search of extragalactic neutrinos. Decay of pions and kaons gives the main contribution to the atmospheric muon and neutrino background (reactions for pion decay are shown in (1.3). The source of the atmospheric background changes with energy. For each parent particle one can calculate the critical energy above which the particle is likely to interact or be slowed down before decaying via leptonic channel. For pions and kaons this energy is 115 and 855 GeV respectively. At energies above several TeV semileptonic decay of very-short lived charmed particles might become a dominant source of the atmospheric background, coming, mainly from the following decay modes:

$$D \rightarrow K + \mu + \nu \tag{1.22}$$

$$\Lambda_c \rightarrow \Lambda_0 + \mu + \nu \tag{1.23}$$

Muon rate versus the overburden distance in the rock is shown in Fig. 1.14. Approximately, one can decrease the muon background by two orders of magnitude for 2 Km of water equivalent depth. However, even at these depths the downgoing atmospheric muon background is still higher than the expected signal. Therefore, the search for cosmic signal concentrates on upgoing events, which corresponds to the neutrinos which have passed through the Earth. Since the atmospheric neutrinos may traverse the Earth and lead to the upgoing tracks in the detector, any signal from the cosmic sources would be inferred by observing a significant statistical excess over the background. The signal-to-noise ratio can be improved by rejecting low-energy neutrino events, as the spectrum of the atmospheric neutrinos is steeper than the expected sources spectrum.

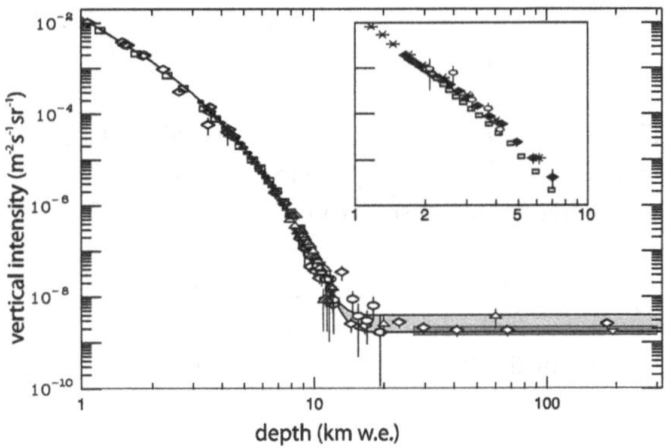

Fig. 1.14 Vertical muon intensity versus depth (1 km water equivalent $= 10^5$ g cm^{-2} of standard rock). The image is taken from Beringer et al. (2012)

1.4 Current and Future High Energy Neutrino Detectors

The largest working Čerenkov neutrino observatory is the IceCube at the South Pole (Fig. 1.15). It has a 1 Gton ice volume equipped with 5160 optical sensors. The sensors are arranged on 86 detector strings. Each string is buried into the ice using hot water drills. Construction began in 2005, when the first IceCube string was deployed and collected enough data to verify that the optical sensors worked correctly. The data collection during the construction period allowed publication of physics results for intermediate detector configurations (IC22 with 22 strings, IC40 with 40 strings and etc.).

Atmospheric muon background blinds the detector in the Southern Hemisphere for the energies below 100 GeV (in the northern hemisphere the Earth acts as a shield against the muons). That is why for the full sky observation the second detector in the Northern Hemisphere is needed. The point sources discovery potential of the ANTARES detector in the Mediterranean Sea is compared with the IceCube discovery potential in Fig. 1.16. The detector uses similar optical sensors and its configuration is described in details in the next chapter. The future KM3NeT at the nearby location will have instrumented volume up to several km^3 and it will be perfectly suited to detect the cosmic neutrino sources in our Galaxy.

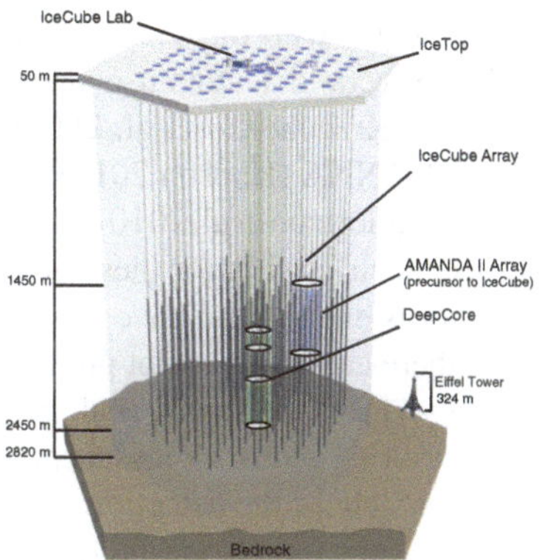

Fig. 1.15 Schematic view of the IceCube detector. The image is taken from Karle (2013)

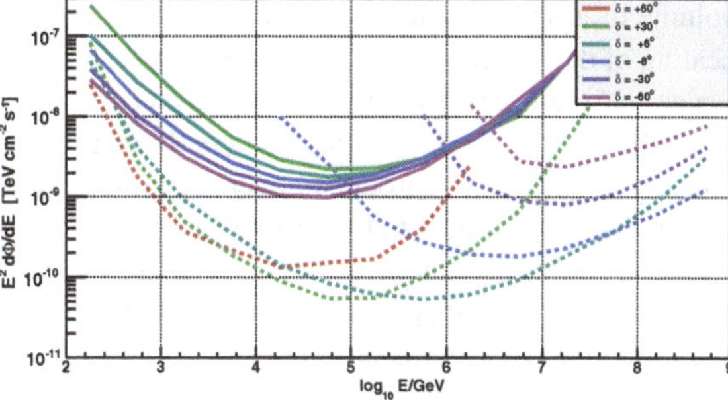

Fig. 1.16 Estimated 5 σ point sources discovery potential of IceCube (*dashed lines*) and ANTARES (*solid lines*) for different declinations. IceCube: 375 days (2008–2009 with IC40 (Spiering 2011); ANTARES: 295 days (2007–2008)

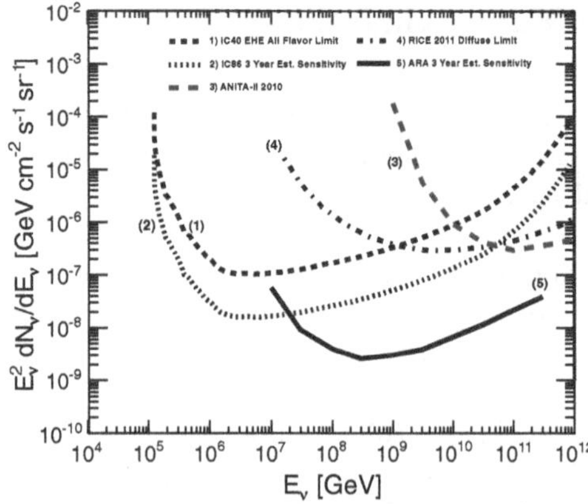

Fig. 1.17 (*1*) IceCube 40, 1-year upper limit to extremely high-energy neutrinos (Abbasi et al. 2011). (*2*) IceCube 86, rough estimate of 3-year sensitivity (Karle 2013). (*3*) ANITA upper limit (Gorham et al. 2012). (*4*) RICE upper limit (Kravchenko et al. 2006). (*5*) the Askaryan Radio Array (ARA) estimated 3-year sensitivity (Allison et al. 2012). Differential limits are corrected for energy binning and flavour differences. The original image is taken from Karle (2013)

At energies above tens of PeV it is more efficient to use radio antennas for neutrino detection. At radio frequencies (MHz–GHz), the emission is coherent and the radiated power scales with the square of the primary particle energy (Alvarez-Muñiz et al. 2012). Several detectors use antennas deployed in ice at Antarctica (existing RICE detector (Kravchenko et al. 2006) and future detectors ARA (Allison et al. 2012) and ARIANNA (Hanson 2013)). Other use antennas on balloons to observe the ice (ANITA (Gorham et al. 2010)). The expected sensitivities and upper limits on the total diffuse neutrino flux for these detectors are compared in Fig. 1.17. The ARIANNA detector is expected to have similar to the ARA detector sensitivity (Hanson 2013). Also there are several pilot projects to observe the radio emission by looking at the Moon surface (see James et al. 2010 for example).

Measuring acoustic pressure pulses in huge underwater acoustic arrays is a promising approach for the detection of cosmic neutrinos with energies exceeding 100 PeV. The pressure signals are produced by the particle cascades that evolve when neutrinos interact with nuclei in water. The resulting energy deposition in a cylindrical volume of a few centimetres in radius and several metres in length leads to a local heating of the medium which is instantaneous with respect to the hydrodynamic time scales. This temperature change induces an expansion or contraction of the medium depending on its volume expansion coefficient. Coherent superposition of the elementary sound waves, produced over the volume of the energy deposition, leads to a propagation within a flat disc-like volume (often referred to as pancake) in the direction perpendicular to the axis of the particle cascade. There are several groups investigating possibility of the acoustic neutrino detection in water (Aguilar et al. 2011) and ice (Descamps 2009).

Sensitivity of the described current high-energy neutrino detectors has almost reached expected cosmic neutrino fluxes. First evidence of the high energy extraterrestrial neutrinos is observed by the IceCube collaboration. No point sources are found so far by the IceCube and the ANTARES collaborations. Future detectors will undertake this task using larger detector volumes and new technologies. For the running detectors there are still a lot of interesting topics which they can manage: extended sources, periodic sources, dark matter and others. Fermi bubbles, to which my thesis is devoted, are one of these promising extended sources for the ANTARES detector.

References

M.G. Aartsen et al., Evidence for high-energy extraterrestrial neutrinos at the icecube detector. Science **342**1242856 (2013a). ISSN 0036–8075, http://www.sciencemag.org/cgi/doi/10.1126/science.1242856

M.G. Aartsen et al., First observation of PeV-energy neutrinos with icecube. Phys. Lett. B **111**(2) (2013b). 021103, ISSN 0031–9007, http://link.aps.org/doi/10.1103/PhysRevLett.111.021103

R. Abbasi et al., Time-integrated searches for point-like sources of neutrinos with the 40-string IceCube detector. ApJ **732**(1), 18 (2011). ISSN 0004–637X, http://stacks.iop.org/0004-637X/732/i=1/a=18?key=crossref.affedca2397066bb2941cfc3db2810e1

J. Abraham et al., Measurement of the energy spectrum of cosmic rays above 1018 eV using the Pierre Auger Observatory. Phys. Lett. B **685**(4–5), 239–246 (2010). ISSN 03702693, http://linkinghub.elsevier.com/retrieve/pii/S0370269310001875

P. Abreu et al., A search for point sources of EeV neutrons. ApJ **760**(2), 148 (2012). ISSN 0004–637X, http://stacks.iop.org/0004-637X/760/i=2/a=148?key=crossref.68e5c9584c4fed5b16100c545a21c89a

M. Ackermann et al., Detection of the characteristic pion-decay signature in supernova remnants. Science **339**(6121), 807–811 (2013). ISSN 1095–9203, http://www.ncbi.nlm.nih.gov/pubmed/23413352

M. Ageron et al., ANTARES: the first undersea neutrino telescope. NIM A **656**(1), 11–38 (2011). ISSN 01689002, http://linkinghub.elsevier.com/retrieve/pii/S0168900211013994

J. Aguilar et al., AMADEUS—the acoustic neutrino detection test system of the ANTARES deep-sea neutrino telescope. NIM A, 626–627 (2011). 128–143, ISSN 01689002, http://linkinghub.elsevier.com/retrieve/pii/S0168900210020772

Q. Ahmad et al., Measurement of the rate of νe + d + p + p + e—interactions produced by B8 solar neutrinos at the sudbury neutrino observatory. Phys. Lett. B **87**(7) (2001). 071301, ISSN 0031–9007, http://link.aps.org/doi/10.1103/PhysRevLett.87.071301

J. Ahrens et al., IceCube Preliminary Design Document, Technical report (2001) http://icecube.wisc.edu/icecube/static/reports/IceCubeDesignDoc.pdf

S. Aiello et al., KM3NeT Technical Design Report, Technical report KM3NeT (2011) http://www.km3net.org/TDR/TDRKM3NeT.pdf

P. Allison et al., Design and initial performance of the Askaryan Radio Array prototype EeV neutrino detector at the South Pole. Astropart. Phys. **35**(7), 457–477 (2012). ISSN 09276505, http://linkinghub.elsevier.com/retrieve/pii/S092765051100209X

R. Aloisio et al., Signatures of the transition from galactic to extragalactic cosmic rays. Phys. Rev. D **77**(2) (2008). 025007, ISSN 1550–7998, http://link.aps.org/doi/10.1103/PhysRevD.77.025007

J. Alvarez-Muñiz et al., Coherent Cherenkov radio pulses from hadronic showers up to EeV energies. Astropart. Phys. **35**(6), 287–299 (2012). ISSN 09276505, http://linkinghub.elsevier.com/retrieve/pii/S0927650511001897

J.N. Bahcall, *Neutrino Astrophysics* (Cambridge University Press, 1989). ISBN 9780521379755

D. Bailey, Monte Carlo tools and analysis methods for understanding the ANTARES experiment and predicting its sensitivity to Dark Matter, Ph.D. thesis (Wolfson College, Oxford, 2002) http://ethos.bl.uk/OrderDetails.do?uin=uk.bl.ethos.249189

I.A. Belolaptikov et al., The Baikal underwater neutrino telescope: design, performance, and first results. Astropart. Phys. **7**(97), 263–282 (1997). http://www.sciencedirect.com/science/article/pii/S0927650597000224

J. Beringer et al., Review of particle physics. Phys. Rev. D **86**(1) (2012). 010001, ISSN 1550–7998, http://link.aps.org/doi/10.1103/PhysRevD.86.010001

S. Bilenky, B. Pontecorvo, Lepton mixing and neutrino oscillations. Phys. Rep. **41**(4), 225–261 (1978). ISSN 03701573, http://linkinghub.elsevier.com/retrieve/pii/0370157378900959

B.T. Cleveland et al., Measurement of the solar electron neutrino flux with the Homestake chlorine detector. ApJ **496**(1982), 505–526 (1998). http://iopscience.iop.org/0004-637X/496/1/505

F. Descamps, Acoustic detection of high energy neutrinos in ice: status and results from the South Pole Acoustic Test Setup, arXiv preprint 0908.3251 (2009), http://arxiv.org/abs/0908.3251

J.A. Formaggio, G.P. Zeller, From eV to EeV: neutrino cross sections across energy scales. Rev. Mod. Phys. **84**(3), 1307–1341 (2012). ISSN 0034–6861, http://link.aps.org/doi/10.1103/RevModPhys.84.1307

P. Gorham et al., Observations of the Askaryan effect in ice. Phys. Lett. B **99**(17) (2007). 171101, ISSN 0031–9007, http://link.aps.org/doi/10.1103/PhysRevLett.99.171101

P.W. Gorham et al., Observational constraints on the ultrahigh energy cosmic neutrino flux from the second flight of the ANITA experiment. Phys. Rev. D **82**(2) (2010). 022004, ISSN 1550–7998, http://link.aps.org/doi/10.1103/PhysRevD.82.022004

P.W. Gorham et al., Erratum: observational constraints on the ultrahigh energy cosmic neutrino flux from the second flight of the ANITA experiment [Phys. Rev. D **82** 022004 (2010)]. Phys. Rev. D **85**(4), 1 (2012). 049901, ISSN 1550–7998, http://link.aps.org/doi/10.1103/PhysRevD.85.049901

J.C. Hanson, The Performance and Initial Results of the ARIANNA Prototype, Ph.D. thesis (University of California, Irvine, 2013) http://arianna.ps.uci.edu/sites/default/files/HansonThesis.pdf

C.W. James et al., LUNASKA experiments using the Australia Telescope Compact Array to search for ultrahigh energy neutrinos and develop technology for the lunar Cherenkov technique. Phys. Rev. D **81**(4) (2010). 042003, ISSN 1550–7998, http://link.aps.org/doi/10.1103/PhysRevD.81.042003

A. Karle, Neutrino astronomy a review of future experiments. Nucl. Phys. B P. Sup. 235–236 (2013). 364–370, ISSN 09205632, http://linkinghub.elsevier.com/retrieve/pii/S0920563213001552

S. Klimushin et al., Precise parametrizations of muon energy losses in water, arXiv preprint hep-ph/0106010 (2001), 1–4, http://arxiv.org/abs/hep-ph/0106010

M. Kowalski et al., Improved cosmological constraints from new, old, and combined supernova data sets. ApJ **686**(2), 749–778 (2008). ISSN 0004–637X, http://stacks.iop.org/0004-637X/686/i=2/a=749

I. Kravchenko et al., RICE limits on the diffuse ultrahigh energy neutrino flux. Phys. Rev. D **73**(8) (2006). 082002, ISSN 1550–7998, http://link.aps.org/doi/10.1103/PhysRevD.73.082002

V. Marin, Charge excess signature in the CODALEMA data. Interpretation with SELFAS2. in Proceedings of the 32nd ICRC, vol. 1, (Beijing, 2011), pp. 291–294, http://www.ihep.ac.cn/english/conference/icrc2011/paper/proc/v1/v1_0942.pdf

M. Markevitch et al., Direct constraints on the dark matter self-interaction cross section from the Merging Galaxy Cluster 1E 065756. ApJ **606**, 819–824 (2004). http://iopscience.iop.org/0004-637X/606/2/819

P. Mészáros, Gamma-ray bursts. Rep. Prog. Phys. **69**(8), 2259–2321 (2006), ISSN 0034–4885, http://iopscience.iop.org/0034-4885/69/8/R01/

F. Reines, C. Cowan, The Reines-Cowan experiments: detecting the poltergeist. Los Alamos Science (25), 4–6 (1997). http://cat.inist.fr/?aModele=afficheN&cpsidt=10450100

C. Spiering, IceCube and KM3NeT: lessons and relations. NIM A, 626–627, S48–S52 (2011), ISSN 01689002, http://linkinghub.elsevier.com/retrieve/pii/S0168900210012465

M. Su et al., Giant gamma-ray bubbles from fermi -lat: active galactic nucleus activity or bipolar galactic wind? ApJ **724**, 1044–1082 (2010). ISSN 0004–637X, http://iopscience.iop.org/0004-637X/724/2/1044/

W. Tucker et al., A search for "failed clusters" of galaxies. ApJ **444**, 532–547 (1995). http://adsabs.harvard.edu/full/1995ApJ..444.532T

F. Vissani, F. Aharonian, Galactic sources of high-energy neutrinos: highlights. NIM A **692**, 5–12 (2012). ISSN 01689002, http://linkinghub.elsevier.com/retrieve/pii/S0168900211023138

Chapter 2
The ANTARES Detector

The ANTARES[1] detector (Ageron et al. 2011) is a pilot project to demonstrate that underwater neutrino telescopes are feasible and that they can operate for several years in the sea water at great depths. The ANTARES detector is located near Toulon, France. The coordinates are $42°50'$ N and $6°10'$ E. Its location makes it sensitive to a large part of the southern sky, including the Galactic Centre region. The detector is located at a depth of 2475 m on the Mediterranean seabed. There is a large plateau in this place (several tens of km^2 with a depth change less than 200 m, Fig. 2.1) which is promising for the further project extension.

In this chapter a general description of the detector design and its operation is given. In particular, characteristics of the used PMTs and the optical modules containing them are described in details. Measurements of the PMT characteristics and simulation of the optical module, which have been part of this thesis work, are essential for the complete detector simulation. The detector simulation itself will be described in Chap. 3.

2.1 Detector Design

The underwater Cherenkov neutrino telescope detector should fulfil the following requirements:

- Due to the very low flux of the high-energy neutrinos from astrophysical sources and to the small neutrino interaction cross-section, a large instrumented volume is required.
- Optimised light detection efficiency and high timing accuracy are needed for the detector elements. The detection units should be capable to detect single photons with a nanosecond precision.

[1] Astronomy with a Neutrino Telescope and Abyss environmental RESearch.

V. Kulikovskiy, *Neutrino Astrophysics with the ANTARES Telescope*,
Springer Theses, DOI 10.1007/978-3-319-20412-3_2

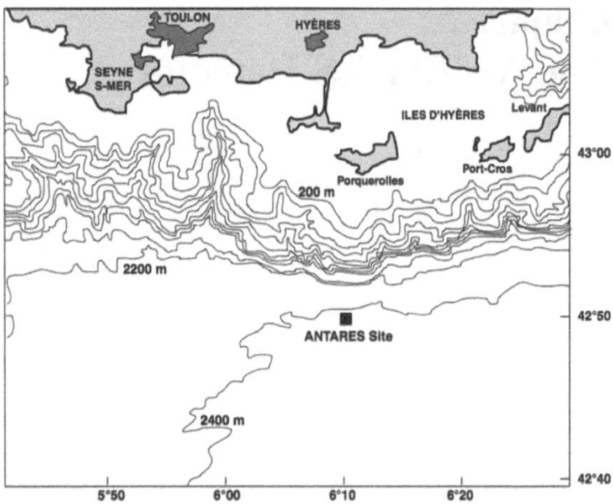

Fig. 2.1 Depth contour plot near the ANTARES site. Each profile line corresponds to 200 m depth change. This image is taken from Palioselitis (2012)

- High pressure resistance. The detector depth of more than 2 Km must be considered to decrease the atmospheric muon rate. The detector components must withstand also additional stress during the installation (shocks, exposure to sunlight etc.).
- The lifetime should be greater than 10 years. The accessibility to the detector is difficult and expensive due to a great depth. Reliability of the components should be high.

To meet these requirements large spherical photomultiplier tubes (10″ radius) were used for the light detection. Their choice is discussed in the next section. For the protection, each PMT and its associated electronics (power base) is housed in a pressure-resistant glass sphere to construct an optical module (OM). There are 885 OMs used in the ANTARES detector.

The arrangement of the OMs in space was optimised by the mean of simulation to have the best neutrino detection efficiency which consists of the effective detection volume and the angular resolution for the tracks. The optimal distances between OMs are correlated with the absorption length in water (maximum is about 60 m as it is described in Sect. 3.3.1). Time coincidence conditions between close OMs (< 1 m) allows to reduce an optical background in the sea water (Sect. 3.3.4). The Fig. 2.2 shows the detector configuration, constituted of OM trios distributed on vertical lines.

The lines are flexible structures attached to the sea bottom by a heavy anchor and kept in tension by a buoy on the top of each line. There are 12 lines with a distance of 60–65 m between them. Any line can be deployed independently from others and released to the surface for maintenance. This simplifies operations at great depth. The lines are connected to the common junction box with 80 Km cable connecting the junction box to the shore station (La Seyne-sur-Mer) where the ANTARES control room is located. Each line (except one[2]) has 25 storeys.

[2]The last line has hydrophones instead of OMs in the five top storeys.

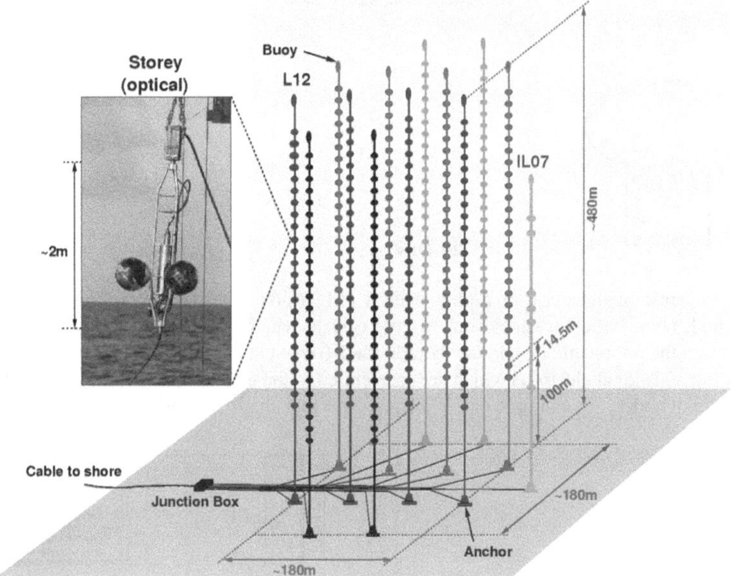

Fig. 2.2 Schematic view of the ANTARES 12-line detector. The image is taken from Palioselitis (2012)

The storey is a mechanical structure to place three OMs to point downwards at 45° with respect to the vertical. OMs are downwards oriented to optimise the upgoing particles detection. Also this orientation allows to protect the OM sensitive area from the sedimentation. The storeys are separated by 14.5 m on the line connected with an electro-optical cable. The first storey is placed 100 m above the anchor. This distance is the minimal distance from the ground where the Čerenkov cone from the up-going particles is developed.

In the following sections the key components of the detector are described in details paying more attention to the PMTs and OMs for which I performed different studies during this thesis.

2.1.1 Photomultipliers

In PMTs arriving photons are converted to electrons via photoelectric effect on the thin metal layer of the so-called photocathode. The electrons are then carried to the stack of the dynode plates with a high voltage between them. Accelerated by the electric field in the tube, the electrons knock out more electrons on each dynode. The gained cascade of the electrons after a stage of several dynodes is sufficient to be seen on the collecting anode even for a single arrived photon. These processes are schematically shown in Fig. 2.3. A typical PMT anode signal due to the single

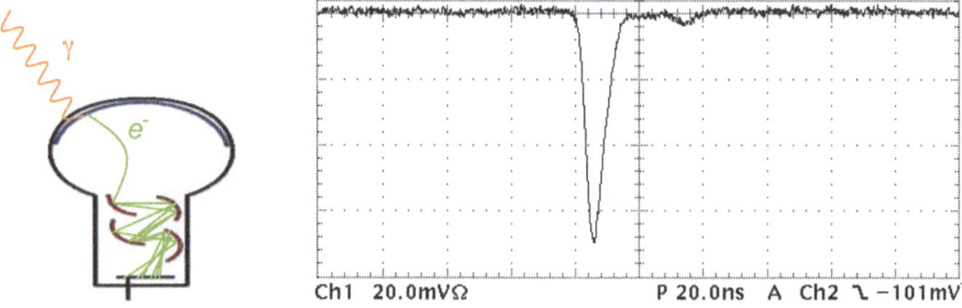

Fig. 2.3 Scheme of detection of a photon by a PMT (*left*). The photon (*yellow*) arrives to the photocathode (*blue*) where it knocks out electron (*green*) which in turn is accelerated in the electric field between the photocathode and the dynode stack (*red*). On each dynode plate there is a chance to knock out additional electrons which are accelerated to the next dynode plate by an electric field between the dynodes. An example of a PMT response to a single photon exposure (*right*)

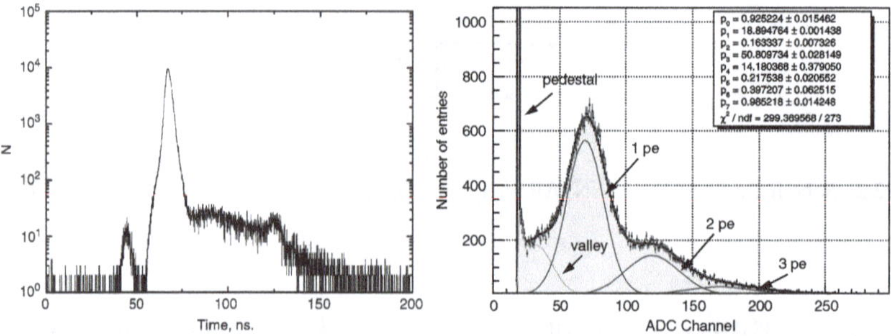

Fig. 2.4 Example of a single photoelectron transit time distribution (*left*). The image is taken from Lubsandorzhiev et al. (2006). Example of a charge distribution expressed in number of digitiser channels, which are proportional to a charge in Coulombs (*right*). This distribution is fitted with several Gaussian functions which correspond to the different number of photoelectrons, emitted from the photocathode. The image is taken from Aguilar et al. (2005)

photon arrival is shown in the same figure. This signal is normally digitised as a time at which the signal amplitude becomes higher than a threshold (typically, 1/3 of the peak) and the total charge integrated in the time window where the signal stays over the threshold (slightly extended to contain a whole signal from a single photon arrived).

A typical time distribution of the detected hits as a response to arriving single photons is shown in Fig. 2.4. The main peak with an approximately Gaussian distribution is seen. The so-called transit time spread (TTS) is referred in following to the full width at half-maximum (FWHM) of this transit time distribution. Additional peaks due to prepulses and delayed pulses are also present. Their origin is discussed in Sect. 2.1.3.

A typical charge distribution for single photons has a Gaussian shape (Fig. 2.4) and it corresponds to the number of electrons gained on the dynode stack (gain). The total charge which corresponds to the mean of this distribution scales linearly for the number of photons simultaneously arrived to the PMT. This makes efficient to express the charge as a number of photoelectrons (p.e.) by dividing the charge by the mean charge value which corresponds to the single photon. Digitised anode signal is called "hit" in following.

Photocathode is typically done using alkali metals and it operates in the vacuum. Probability of a photon to produce a photoelectron is called quantum efficiency and it is about 20–40 % for modern photocathodes. The photocathode should be negatively charged. Electron emission happens due to the photoelectric effect and the electrons escape the photocathode on the opposite site from the arrived photon and then they are driven away by the initial kinetic energy and the electric field.

In modern experiments PMTs are widely used thanks to the high amplitude output (comparing to photodiodes), sensitivity to single photons, linear response to different amount of arrived photons and good timing characteristics. A large variety of PMTs with different sizes and shapes of photocathode are available commercially. PMTs with a spherical photocathode have the best timing resolution, as the photoelectrons cover almost the same distance to the first dynode from any point on the photocathode.

PMTs with a large photocathode are more sensitive to the magnetic fields (for example, the Earth's magnetic field) as the electrons are more deflected on greater distances. The magnetic shielding is important for such PMTs. PMTs have vacuum inside the bulb to have a better electron collection efficiency and reduce the spurious pulses arriving from the ionised gas. Together with the thin glass in front of the photocathode this makes them a very fragile device. Photocathode could be damaged by a high light flux and also photocathode material can degrade with a time (ageing).

2.1.2 PMT Choice

The PMT model was chosen during research and development phase of the ANTARES project (Aguilar et al. 2005). Several requirements for the PMT choice should be satisfied:

Size The previously mentioned requirements of the detector size can be satisfied if the total density of the photomultipliers is low (10^{-4} PMT/m^3) to reduce the cost of mechanics. These in turn leads to the fact that the number of detectable photons produced by muons is very low for each detection unit. This requires a large detection area for each unit. Simulation shows that the geometrical area of the PMT should be more than 400 cm^2 if the photocathode quantum efficiency $\geq$20 % is considered.

Timing The time resolution has an impact on the reconstructed muon angular resolution. The ANTARES experiment aimed absolute time calibration better than 1 ns. The timing uncertainty due to the jitter of the PMT response should also remain of the order of 1 ns, in order not to degrade significantly the angular resolution. It was decided that TTS should not exceed 3 ns (Aguilar et al. 2005).

Gain The noise level of the electronic circuit used to digitise PMT pulses requires a single photoelectron (SPE) amplitude to be of the order of 50 mV on a 50 Ω load (Feinstein 2003). This requires the photomultiplier gain to be at least 5×10^7. For the absolute calibration of the telescope, it is necessary to be able to properly isolate a single photoelectron from the pedestal. This requires peak-to-valley ratio (P/V) to be at least 2, as computed from the charge spectrum. The charge resolution should be less than 40 %.

Noise The optical background in the sea water at ANTARES site is about 60 kHz for a 10″ PMT (see Sect. 3.3.4). The dark count rate should be negligible comparing to this rate so the maximum dark count rate <10 kHz is required. The dark count rate includes the PMT noise (spontaneous electron emission from a photocathode and dynodes and other processes) and the radioactive decay in the glass and the optical gel of the OM (^{40}K decay mainly).
A signal from PMTs can be affected by different spurious pulses. Basically one distinguishes prepulses, which arrive before the main pulse together or instead of it, delayed pulses which arrive with a 10–100 ns delay instead of the main pulse, afterpulses which arrive together with the main pulse in several time windows: 10–100 ns (afterpulses type 1) or 100 ns–16 μs (afterpulses type 2). The detailed study with the description of the physical process involved for these pulses is done in Sect. 2.1.3.

A large variety of PMT models have been tested and characterised taking into account the constraints listed above. At the end of the tests, the Hamamatsu R7081-20 model was chosen for the ANTARES detector. The full sample of the delivered PMTs (912) has been tested with a dedicated test bench in order to calibrate the sensors and to check the compliance with the specifications. The number of rejected tubes was low (17, their peak/valley ratio being too low), these tubes were replaced by the manufacturer (Ageron et al. 2011).

The ageing was studied with a test setup which simulates the optical background in the sea. The total charge of 500 C, corresponding to 10 years in the sea was collected by each of the three tested PMTs. The LED's pulsers are programmed to emit light continuously with a rate of 500 kHz and bursts of 5 MHz lasting 0.1 s every second to simulate extreme conditions of the optical background in sea water (discussed in details in Sect. 3.3.4). Over the first 100 days, all three PMTs showed an increase of 50–70 % in gain, before they level off. This phenomenon corresponds to the running-in phase, rather than the ageing. All other important energy and timing properties of the PMTs (P/V, energy resolution and TTS) were stable over the test period. The dark count rate remained also stable with time. However, the afterpulses type 2 rate showed an increase (around 30 %) during the test, possibly caused by the fact that these pulses are usually due to the ionised gas, which quantity increases with time (Aguilar et al. 2005).

The selected PMTs fulfil the requirements. For my thesis I performed several measurements with one of these PMTs to get acknowledged with the principles of PMT work and to insert measured data to the simulation of the ANTARES detector.

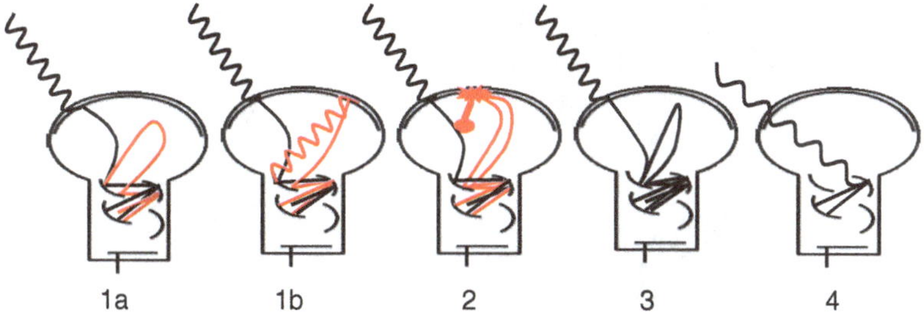

Fig. 2.5 Afterpulse type 1 (*1a, 1b*), afterpulse type 2 (*2*), late pulse (*3*) and prepulse (*4*) origins in a PMT

2.1.3 PMT Characteristics Measurement

Since the track reconstruction is based on the light arrival to the PMT it is necessary to know its transit time spread precisely to be used in the detector simulations later. This transit time spread is affected by several processes. Also, in addition to the normally detected photon signal, correlated pulses may appear with some delay. The spurious pulses affecting TTS or appearing after the normal pulse may be summarised as following (Fig. 2.5):

1. **Afterpulses type 1**: they appear in the first 80 ns after the main pulse and can be produced by the luminous reaction on the dynodes while bombarded by electrons or by the electrons escaped from the dynode in the direction to the photocathode.
2. **Afterpulses type 2**: appear in 80 ns–16 μs interval after the main pulse due to the ionisation of the phototube gas by the photoelectron. Different ions give different contributions to the afterpulses time distribution (H^+, He^+ and heavy ions CH_4^+, O^+, N_2^+, O_2^+, Ar^+ and CO_2^+).
3. **Late pulses**: a primary photoelectron suffers from the elastic or inelastic scattering in the first dynode without a secondary electrons emission, it turns towards the photocathode making a loop and only after it creates a cascade of electrons in the dynodes. The arrival time of the hit will be delayed.
4. **Prepulses**: they appear due to the direct photoelectron emission on the dynodes from the photons which passed the photocathode without an interaction. Due to this, the main pulse arrives earlier and with a smaller charge.

In order to study these processes the following test setup was developed. At first, a bare R7081-20 10″ PMT with the same power base as in the ANTARES detector was used. The PMT was located in a black light tight box with a feed through for power supply and anode signal output. In the same box a Hamamatsu Picosecond Light Pulser PLP10-044C was located. This laser had a wavelength of 440 nm, a pulse width of few picoseconds and a tunable frequency from 1 Hz to 1 MHz. The beam output was connected via an optical fibre to an attenuator and then to a collimator to illuminate the PMT photocathode as shown in Fig. 2.6. A compact diffuser obtained sanding a thin plexiglas sheet ensured not a perfect but an acceptable uniform illumination

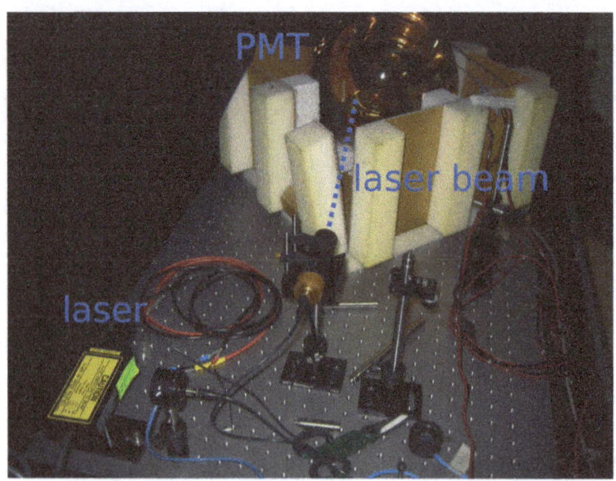

Fig. 2.6 Photograph of the experimental setup used for the ANTARES PMT studies

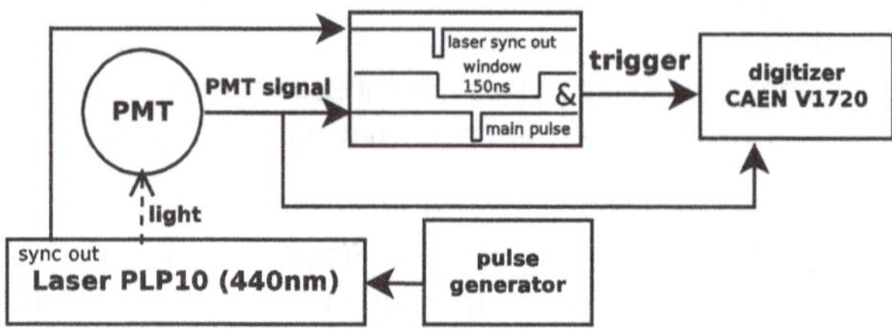

Fig. 2.7 Scheme of the acquisition system used for the ANTARES PMT studies

of the surface. The light from the laser was tuned to obtain, at the photocathode window, $\approx 10^{-2}$ photons/pulse to produce on average one signal at the PMT anode every 500 light pulses sent by the laser. In these conditions one can assume that the laser is working in the single photoelectron (SPE) mode. The data acquisition system was extremely simple. The output of the PMT anode was split to a 0.3 p.e. threshold discriminator and to a fast, 1 GHz, CAEN digitiser V1731. The output of the discriminator was stretched to 150 ns and put in coincidence with the Laser SYNC out as shown in Fig. 2.7. In order to exploit the full timing capabilities of the CAEN fast ADC, both the coincidence and the anode signals were sent respectively to channel 0 and 2 of the digitiser. Worth to notice that the logic unit output was always synchronous with respect to the laser SYNC out: the arrival time of the PMT signal was therefore measured relative to the light pulse. In practice, the coincidence NIM signal in channel 0 was used to trigger the acquisition in the following 16 μs time window. To check that this system provided the correct timing, I have sent the two outputs of the same discriminator, one delayed by a fixed quantity, to the CAEN digitiser: the obtained distribution showed a sharp peak, 1 ns FWHM, as expected. The example of the digitised pulse is shown in Fig. 2.8.

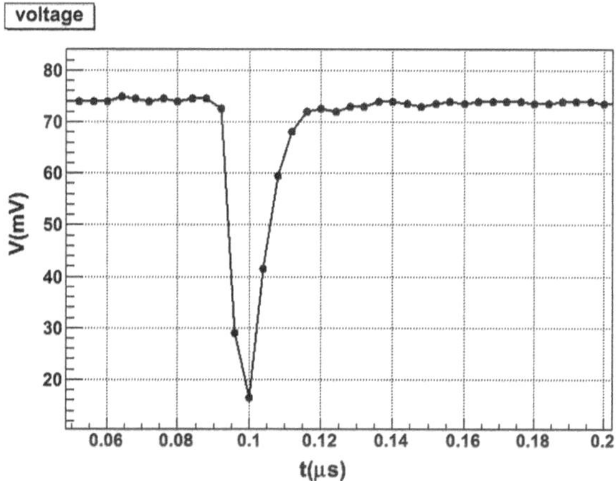

Fig. 2.8 The anode pulse as measured by the CAEN digitiser with the linear interpolation which is used for the charge calculation

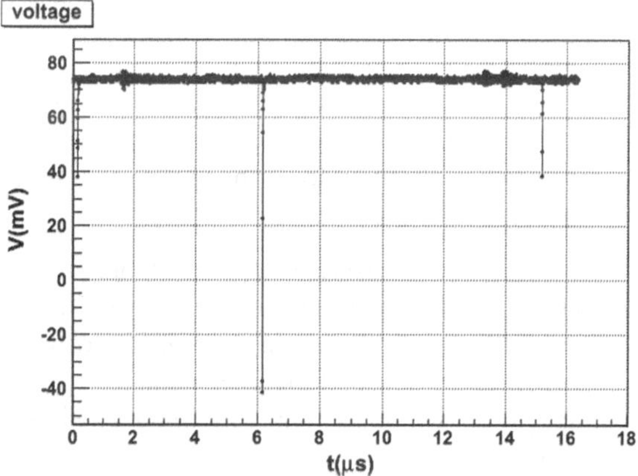

Fig. 2.9 The PMT anode output together with possible after-pulses at 6 and 15 μs from the laser pulse

A typical digitised PMT anode output is plotted in Fig. 2.9 in the 16 μs time interval. Together with the main pulse at the origin of the time axis, after-pulses are seen after 6 and 15 μs. Also some electronic noise can be seen at $\sim$2 and $\sim$13 μs. Sampling rate was 1 GHz and voltage was digitised in 4 mV steps. The time and charge analysis consisted of the next steps:

1. Determination of DAQ offset in the 16 μs interval and subtract it from the signal.
2. Search for all peaks and their charge by iterating time sample by time sample. When the signal is higher than 10 mV, corresponding to a threshold of about 0.2–0.3 p.e., the program looks for the presence of a peak and finds the left and right boundaries of the peak T_l and T_r where the amplitude of the signal is higher than 10 mV. The time of the maximum amplitude is considered as the time of the pulse. Two peaks within a 25 ns time interval are assigned to the same event and the time of the event is the time of the maximum of the two peaks. The charge is

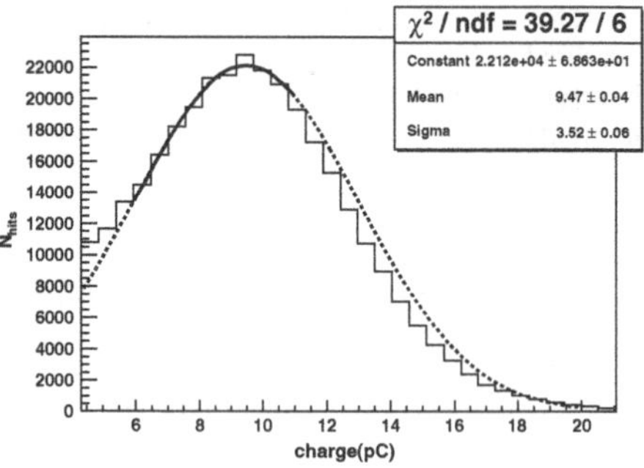

Fig. 2.10 Measured ANTARES PMT charge due to a single photon arrival

evaluated integrating the peak within the Simpson approximation from $T_l - 5$ to $T_r + 10$ ns.

3. The hits in the first 200 ns of the time window are considered as "normal" (not afterpulses). The plot of the corresponding charge distribution is shown in Fig. 2.10. The plot is well fitted with a single Gaussian distribution to prove that contributions of two or more photo electrons to the main hits are negligible. The left part at low charge is a little bit distorted due to the contribution of prepulses with low charge. The mean value of the charge of the 1 p.e. peak $\tilde{q}$ is found from the fit. After the fit is done, the charge of a hit q is divided by $\tilde{q}$ to obtain the charge in photoelectrons.

The performed measurements of TTS are shown in Fig. 2.11. The peak of the normal pulses from the laser photons is clearly evident at t ~100 ns while random background forms a plateau in accordance to the dark current measured rate of ~1.6 kHz in the absence of the laser signal.

The first measurement was done with a uniform diffuse illumination and the second was done with a point like beam towards the centre of a PMT photocathode. In both cases a significant tail towards the increasing arrival times can be observed. This tail might be ascribed to an inelastic scattering of the photoelectron in the first dynode (Lubsandorzhiev et al. 2006). It is expected that this effect can be a bit larger for the uniform illumination. Also there is an additional structure at ~50 ns from the main peak which is ascribed to the elastic scattering of the photoelectron. In the latter case the momentum of the scattered electron is fixed which creates a fixed delay as well. To this extent, assuming a uniform electric field of ~60 V/cm at 14 cm distance between the first dynode and the photocathode, the 50 ns of the peak roughly corresponds to the "round trip" of an electron from the first dynode to the photocathode and its way back. The contribution of the late pulses as deduced from the plot are about 2.5 % in the "elastic" peak.

On the left side from the main peak, the pre-pulses contribution may be seen in the 10–15 ns preceding the main pulse. As expected, the contribution of the afterpulses is larger for the point like illumination which is directed toward the first dynode. In the same plot the dashed curve represents the Gaussian fit of the left side from the main peak.

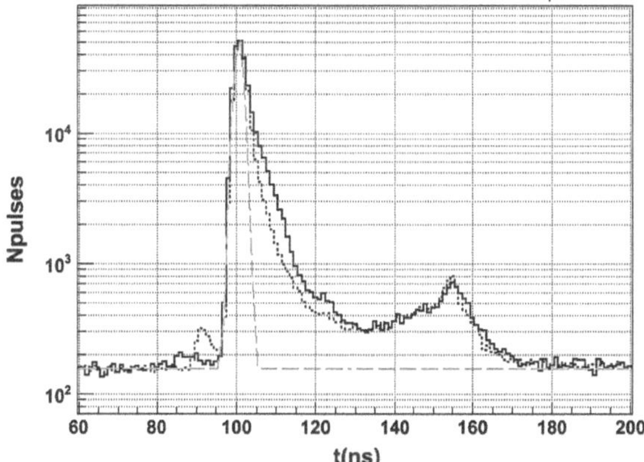

Fig. 2.11 Photon time arrival (transit time distribution) for the measurements in a dark room with the different sources: point-like illumination with laser (*dotted curve*), uniform illumination with laser plus diffuser (*continuous curve*). Delayed pulses with elastic peak at 145 ns are seen for both measurements. Prepulses at 95 ns are seen better for the point-like illumination. Gaussian fit of the left part of the spread functions is also shown (*grey dashed*)

In Fig. 2.12 the results from the measurements on the afterpulses are presented. One can see hits arriving shortly after the main hit (in 10–40 ns) or with a longer delay (6–8 μs). The total contributions of these afterpulses is 1.0 % and 0.8 % correspondingly. The late afterpulses (type 2) do not affect the track reconstruction of passing muons in the ANTARES detector but simply increase the light background. The fast afterpulses, instead, should be considered for the reconstruction algorithms. Also it can be seen that there is an increase of the afterpulses with a higher charge in respect to the charge distribution of single photoelectron hits (Fig. 2.10).

The obtained results are comparable with the measurements done during the PMT choice study (Aguilar et al. 2005). Similar measurements for the PMT used by the NEMO collaboration[3] (R7081) and the high quantum efficiency prototype of the same PMT are done during this thesis (Aiello et al. 2011).

2.1.4 *Optical Module*

The scheme of the optical module is shown in Fig. 2.13. It contains a PMT with its protection from the high pressure (glass sphere) and magnetic fields (metal cage) together with the PMT electronics and a calibration LED. The PMT is attached to the glass sphere via optical gel to reduce the unwanted reflection.

The protective envelope of PMT is a glass sphere of the kind routinely used by sea scientists for the buoyancy and for the instrument housing. These spheres, thanks to

[3]Now it is part of the KM3NeT project.

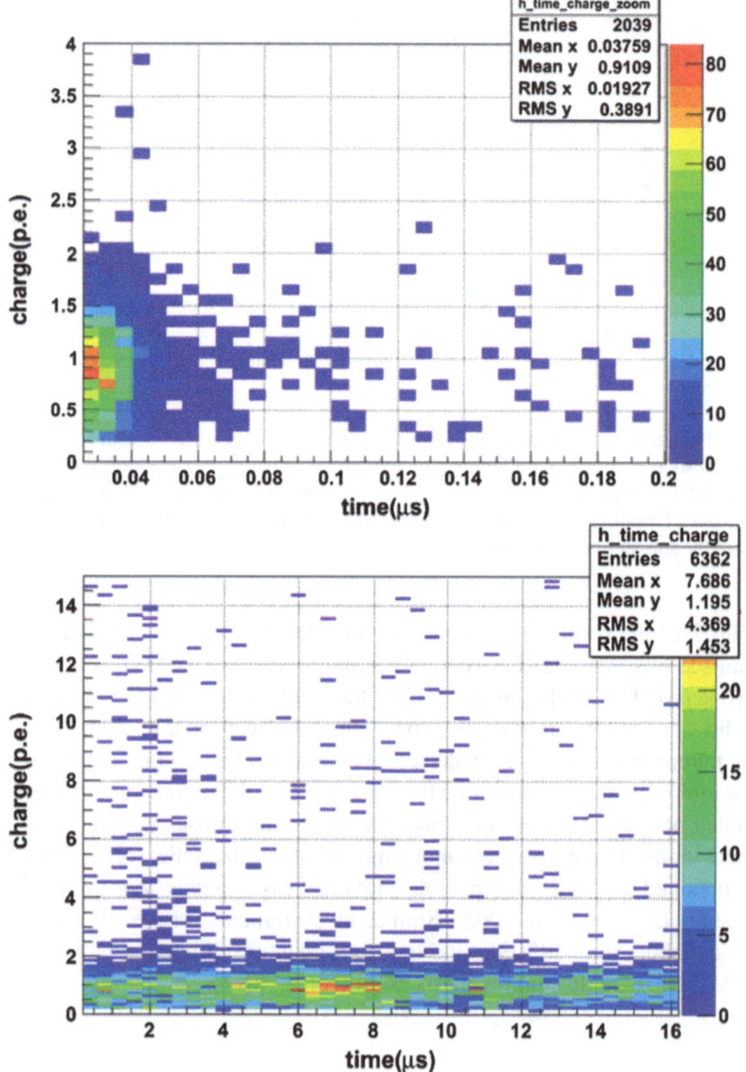

Fig. 2.12 Time versus amplitude distribution of afterpulses type 1 (*top*) and type 2 (*bottom*)

their mechanical resistance to a compressive stress and to their transparency, provide a convenient housing for photodetectors. Commercially available Vitrovex®17″ glass spheres were used.[4] The pressure qualification test allows the sphere usage till the depths of 7000 m.

[4]Nautilus Marine Service GmbH, http://www.nautilus-gmbh.de.

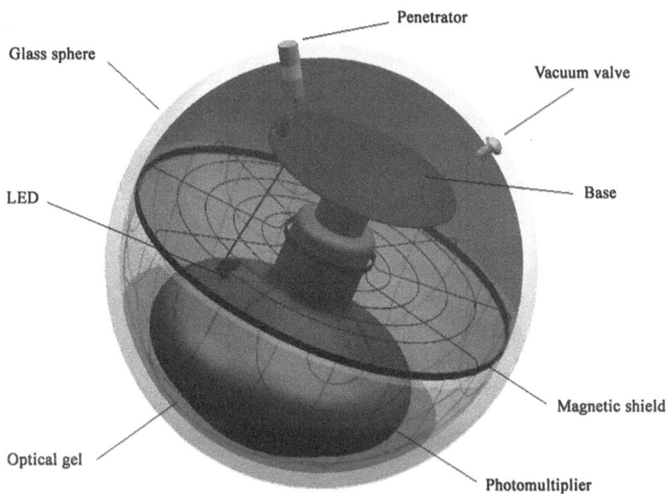

Fig. 2.13 Schematic view of the ANTARES OM and its components. The image is taken from Amram et al. (2002)

The PMT must be protected from the Earth's magnetic field which can deflect the electrons inside the tube and both reduce the PMT efficiency and affect the timing characteristics. The shielding was done with a hemispherical grid made of wires of μ-metal[5] closed by a flat grid on the rear of the bulb. The efficiency of the screening becomes larger as the size of the mesh is reduced and/or the wire diameter is increased, however the drawback is a shadowing effect on the photocathode. The compromise adopted by the ANTARES collaboration, a mesh of 68 mm^2 and wire diameter of 1.08 mm, results in a shadowing of less than 4 % of the photocathode area while reducing the magnetic field by a factor of three. Measurements performed in the laboratory show that this shielding provides a reduction of 0.5 ns on the TTS and a 7 % increase on the collected charge with respect to a naked, uniformly illuminated PMT (Ageron et al. 2011). The μ-metal cage is immersed in the gel and hold by it in the OM.

Instead of the passive PMT base provided by Hamamatsu, a new base (a modified version of the PHQ5912 base developed by ISEG[6]) was used. The high voltage and the distribution for the dynodes is generated by the base itself from the input 48 V used in ANTARES. Additional modification of the base were done to reduce the anode signal reflection.

On the rear part of the bulb of the PMT, a blue LED is glued in such a way to illuminate the pole of the photocathode through the aluminium coating. This LED is used to monitor the internal timing of the OM. A complete description of the ANTARES OM may be found in Amram et al. (2002).

---

[5]Sprint Metal, Ugitech, http://www.ugitech.com.

[6]ISEG Spezialelektronik GmbH, http://www.iseg-hv.com.

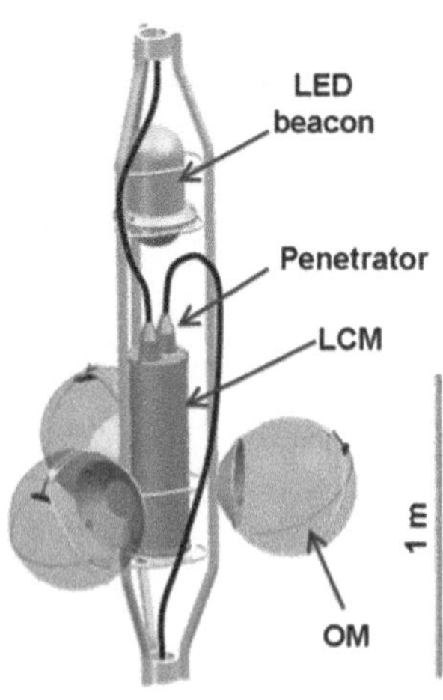

Fig. 2.14 Drawing of a storey equipped with 3 OMs which are connected to a local control module at the centre of the storey. The image is taken from Ageron et al. (2011)

2.1.5 Storey

Each OM is connected to a storey frame using a support system done with a titanium (Ti) wires, plate and bolts (Fig. 2.14). The storey frame hosts electronics for OM signal digitisation and transmission—Local Control Module (LCM). The LCM is placed in the titanium cylindrical container in the centre of the frame. A standard LCM contains the following elements:

Local Power Box Provides the 48 V for the optical modules and several different low voltages for the electronics boards. An embedded micro-controller allows the monitoring of the voltages, the temperatures and the current consumptions as well as the remote setting of the 48 V for the OMs. It is fed by the 400 V DC from the bottom of the line.

Clock The clock reference signal coming from shore reaches the bottom of the line where it is repeated and sent to each sector. Within a sector, the clock signal is daisy-chained between LCMs. The role of the clock card is to receive the clock signal from the lower LCM, to distribute it within the boards of the storey and to repeat it toward the upper LCM of the sector.

ARS boards The ARS motherboards host the front-end electronics of the OMs (one board per OM). This front-end electronics consists of a custom-built Analogue Ring Sampler (ARS) chip (Feinstein 2003) which digitises the charge and the time of the analogue signal coming from the PMTs.

DAQ/SC The Data Acquisition/Slow-Control card host the local processor and memory which are used to handle the data from the ARS chips and from the slow control, respectively. The processor has a fast Ethernet controller (100 Mb/s) that is optically connected to an Ethernet switch in the MLCM of the corresponding sector. The specific hardware for the readout of the ARS chips and data formatting is implemented in a high density field programmable gate array (FPGA).[7] The data are temporarily stored in a high capacity memory (64 MB SDRAM) allowing a de-randomisation of the data flow.

Compass board The compass motherboard hosts a TCM[8] sensor which provides heading, pitch and roll of the LCM. Those parameters are used for the reconstruction of the line shape and PMT positions.

Due to the segmentation of a line in sectors, one in five LCMs, called Master LCM (MLCM), acts as a master for other LCMs of the same sector and houses additional boards to collect electrical signals from the LCMs and convert them to the optical signal distributed to the bottom of the line.

The acoustic neutrino detection is integrated into ANTARES. The acoustic storeys are modified standard storeys with the PMTs replaced by acoustic sensors with custom-designed electronics for signal processing. The test system consists of six acoustic storeys with six acoustic sensors each. They are located on the last line of the detector and one additional instrumentation line. The detailed description may be found here (Aguilar et al. 2011a).

Several storeys also host light sources (LED optical beacons) which are used for time calibration (Sect. 2.3.1).

2.1.6 Line Infrastructure

Storeys are connected with a EMC which has three roles:

- optical data link: 21 single mode optical fibres
- power distribution: 9 electrical conductors
- mechanical link: breaking tension above 177 kN.

When subjected to a uniform horizontal sea current, as present at the ANTARES site, the 3-fold periodic symmetry of the storey induces a torque which is a function of the actual azimuth of the storey. The storey is in stable equilibrium when one of the three OMs is upstream of the current. Between two adjacent storeys, the EMC acts as a torsion spring tending to keep them at the same relative angle.

Each string is attached to its own Bottom String Structure (BSS). The function of the BSS is to anchor the line to the seabed with the capability of a recovery. The BSS is made of two parts: an unrecoverable dead weight laid on the seabed and a recoverable part sitting on top. The two parts are connected by a release system remotely controlled by acoustic signals. The dead weight is a horizontal square plate made of 50 mm thick carbon steel. The line stability requires a dead weight of 1270 kg in water, which means 1500 kg of steel.

The line is linked to the junction box by the interlink cable (IL), an electro-optical cable laying on the seabed and connected to the line at the level of the BSS by an

[7] Virtex-EXCV1000E, http://www.xilinx.com.

[8] PNI Sensor Corp., http://www.pnicorp.com.

underwater mateable connector, developed by the ODI company.[9] The remote line release implies an automatic disconnection system for the IL cable: the plug of the IL is fixed to the dead weight while the socket is located on the recoverable part of the BSS, at the end of a pivoting arm in such a way that it is extracted from the plug at the beginning of the ascent.

The BSS holds two lithium batteries powered transponders[10] in titanium cylinders which are equipped with release mechanism. The acoustic beacon capability of the transponders is employed in the Low Frequency Long Baseline (LFLBL) positioning and navigation system to monitor the position of the line anchor during its deployment and to determine its geodetic location with a precision of $\approx$1 m.

In addition to the release system, the BSS holds:

- a 1.8 m long titanium container housing the power module and the control electronics of the line
- a high resolution pressure meter, used for line positioning
- an acoustic transceiver at the top of a 3.6 m long rod of glass-epoxy, used as a reference emitter for the High Frequency Long Base Line (HFLBL) positioning system
- optional sound velocimeter, laser beacon, seismometer depending on the line
- a weight to keep the line vertical and under enough tension after release, even when the buoy reaches the sea surface.

The electronics container is the assembly of two cylinders similar to those of the LCMs. The power module part, at the top, contains transformers which deliver the five 400 V DC supplies needed for the sectors, starting from the 480 V AC provided by the junction box. An embedded micro-controller operates the remote powering of sectors. Voltages, temperatures and current consumptions are monitored. The micro-controller can also detect anomalies and is programmed to turn off the power in case of over-consumption. Like a standard LCM container, the control electronics cylinder houses a crate equipped with compass board, clock board and DAQ cards. There are boards which receives 20 MHz clock signal via an optical link from the shore and distribute it as electrical signal to the sectors. Also, the crate is equipped with a board which merges the optical signals from the sectors to distribute them through the single optical line. Each signal is sent to the shore by using a different laser colour.

2.1.7 Detector Infrastructure

The detector lines are connected with a Junction Box (JB) via interlink cables which is in turn connected to the shore using the main electro-optical cable. The JB is a

[9]Ocean Design Inc. (ODI), http://www.odi.com.

[10]Type RT 861 B2T; IXSEA/Oceano, http://www.ixsea.com.

pressure resistant titanium container mounted to the JB frame. The JB and its frame provide the following facilities:

- connection of the main electro optical cable and of the sea return power electrode
- power transformer housing
- line over-current protection system
- remote diagnostic system
- 16 electro-optical sockets to plug the interlink cables

The main electro optical cable (MEOC)[11] has been deployed from the site to the shore by a specialised cable-laying ship and crew under the responsibility of Alcatel. A standard undersea cable configuration with a single conductor was chosen to minimise the cable cost and weight. The use of the sea for current return reduces ohmic losses by a factor 4 compared with a cable sharing equivalent cross-section between supply and return conductors. The cable contains 48 monomode optical fibres in a stainless steel tube surrounded by "pressure vault" of two windings of steel armour wires. A tubular copper power conductor surrounds the vault and delivers current up to a maximum of 10 A to the junction box.

The onshore infrastructure consists of two separate buildings, the Shore Station housing control and data management infrastructure and providing space for onsite personnel, and the Power Hut devoted to power distribution requirements. The shore station is situated at La Seyne-sur-Mer. The building has three rooms dedicated to the operation of the ANTARES experiment: a computer room, a control room, and a service room. Computer room hosts racks for the clock crate, crates for the optical signal separation and the PC farm for data filtering and storage. The control room contains various computers for apparatus control and status monitoring.

The Power Hut is located near to MEOC landing point at Les Sablettes, and is connected to the latter with a fibre optical link 1.5 Km long. The Power Hut is connected to the 60 kVA, 400 V three-phase electrical distribution from Électricité de France. The building has been adapted to house the transformer 400/4000 V, the MEOC to shore link rack as well as the return current electrodes.

2.2 Data Acquisition

Data acquisition system (DAQ) (Aguilar et al. 2007) of ANTARES serves to digitise analogue signal of PMTs, transfer the data to shore, filter it and store it on the disks. The main concept chosen is "all data to shore". According to it the underwater part of the detector is performing only digitisation and all the filtering is done off-shore.

The presence of ^{40}K in the water and the bioluminescence introduces a lot of background signal which is not possible to store as raw data on disks for the long duration. For the first step the protection in the detector data filters against the bioluminescence bursts known as a high rate veto (HRV) is implemented. If a rate of a PMT exceeds

[11] Alcatel URC3 Type 4 (unrepeatered); Alcatel-Lucent, http://www.alcatel-lucent.com.

HRV value which is 400 kHz by default then the data from this PMT are not used. The same cutoff occurs when a PMT registers an excessive counting rate for a long period, overfilling the buffer (X_{off}). Lately, the filtering is done with triggers which have different purity and efficiency on the physics events (usually muons) detection. The filtered data is saved later on tapes which are accessible through the internet for any ANTARES participant.

2.2.1 Signal Digitisation and Data Transmission

A main digitisation method consists of the measurement of time at which negative anode signal passes a threshold (falling edge) and a charge integration over some time after the threshold. The threshold is typically set to 1/3 of the average peak signal and the time is fixed to 25 ns (which covers a length of most 1 p.e. signals). A combined charge-time information is called a level 0 (L0) hit. The signal is read-out and digitised by a custom application-specific integrated circuit, the analogue ring sampler (ARS) (Feinstein 2003). A local clock in ARS is used for the determination of the hit time arrival. The time resolution of the system is better than 0.4 ns. An ARS after digitisation of a signal has an about 225 ns dead time. Each PMT is connected to a pair of ARS readouts which are activated by a token passing through the token ring between these ARS. It takes about 13 ns to pass the token so during this time the PMT signal is not elaborated. The timeline of this process is shown in Fig. 2.15.

Three PMTs have in total 6 ARS which are connected to a FPGA placed in the storeys local control module (LCM). FPGA arranges hits produced by each ARS in a time window of 104.8 μs to a separate data frame. A 20 MHz clock generated off shore is used to provide a common time for each ARS. These clocks are synchronised with a master clock through the optical fibre network. A scheme of these LCM components is shown in Fig. 2.16. The master clock is in turn synchronised to the GPS time with an accuracy of ~100 ns (Aguilar et al. 2011b). This time affects only the absolute neutrino arrival time and it is negligible in respect to the Earth rotation period.

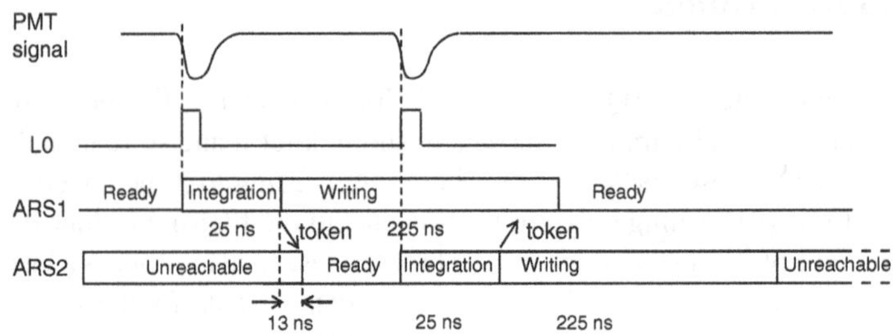

Fig. 2.15 Timeline of two consecutive PMT pulses and the corresponding occupancy in the front-end electronics. The image is taken from Aguilar et al. (2010)

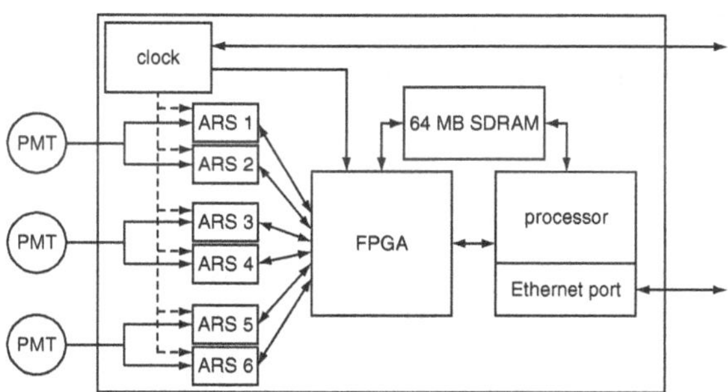

Fig. 2.16 Schematic picture of the main hardware components in the electronics module of a storey Aguilar et al. (2007)

Transmission Control Protocol and Internet Protocol (TCP/IP) together with ControlHost package is used for communication and data transfer (Gurin and Maslennikov 1995). The ControlHost requires a server which is placed in the control room and clients which send and receive the data through the server to another clients using tags. Each LCM has a processor which runs vxWorks operating system[12] with a DaqHarness program which sends the data, collected from each ARS to a ControlHost server. The server distributes the data to the DataFilter clients which group the data by the time organising the data time slices of 104.8 μs from the data frames. Once a time slice is ready, it is processed by its DataFilter with a trigger software. The filtered data is sent in turn to the ControlHost server which forwards it to a DataWriter process that writes it to the disk. At the end the data from the local discs is transferred to the Lyon computing centre through the Internet and stored on tapes.

2.2.2 Data Filtering and Storage

All the digitised signals from PMTs are transferred to the shore. The total data output is about 0.3–0.5 GB/s during low bioluminescence periods. This high data flow contains mostly the optical background and so it needs to be filtered before the storage. The filtering is done with different trigger algorithms which are designed to detect signal from particles in the detector by the space-time correlation between hits (de Jong 2005). These trigger algorithms are implemented in the DataFilter program which triggers PMT hits and produces Physics Events containing all data in a window (-2.2, 2.2 μs) from the triggered hits. This time window covers well a muon passage time through the detector.

All the collected hits are called level 0 hits (L0) hits in ANTARES detector. They are mostly coming from the optical noise. The first level cut searches for coincidences between the hits in the LCM between the 3 PMTs which are very close to each other. It assumes that the Čerenkov light coming from muons should hit the nearby PMTs. A distance between PMT centres in LCM is about 0.7 m. The time window chosen is

[12]http://www.windriver.com.

20 ns which is much more than sufficient to detect scattered hits. The rate of so-called L1 hits per PMTs pair can be estimated as:

$$f_{\mathrm{L1}} = 2\tau f_{\mathrm{L0}}^2 \sim 2 \times 20 \times 10^{-9} \times (100 \times 10^3)^2 \sim 400\ \mathrm{Hz}, \tag{2.1}$$

assuming $f_{\mathrm{L0}} = 100\,\mathrm{kHz}$ L0 rate per PMT. The number of PMT in LCM is three so the number of pairs is also three. This means the rate of the L1 hits is at least two orders of magnitude lower than the L0 rate in the detector. Additionally, the single hits with a charge of more than 3 p.e. are added to the L1 hits. The background in these hits is mainly due to afterpulses which rate is about 0.18 % (from Fig. 2.12).

The second level triggers are applied mostly on L1 hits searching clusters (L2 hits) and checking the compatibility with different track directions. Clusters are formed by the hits connected with a so-called causality relation, which is the following for the two hits (t_i, r_i) and (t_j, r_j):

$$|t_i - t_j| \leq |\bar{r}_i - \bar{r}_j|/v_g + \tau, \tag{2.2}$$

where $v_g = c/n$ is the group velocity of light in the water and τ takes to account a dispersion and PMTs position uncertainties. There are several triggers developed by the ANTARES collaboration. They have different efficiency on the neutrino detection and a different event purity. The triggers which are used for the data taking are set for each run separately depending on the optical background conditions. The main strategy is to have a final detector event rate of order of 100 Hz which produces about 1 GB of physics events per 1–2 h of data taking.

2.3 Detector Calibration

The detector calibrations are important for the event reconstruction precision. The calibration details are written to the database as a calibration set. There are two types of sets—sets which are used for the online detector work (these are needed for the triggering) and more precise offline calibrations are using the data already collected (these are used for the event reconstruction).

2.3.1 Time Calibration

The time calibration in ANTARES is described in details in Aguilar et al. (2011b). Before the deployment of the detector strings the preliminary time calibration is done in laboratory. After the deployment an optical beacons system (OB) is used for the calibration adjustments (Ageron et al. 2007). The four LED optical beacons which emit the blue light are placed in each detector line on the sectors 2, 9, 15 and 21 (counted from the bottom to the top). Each of the OB has its small PMT close to the LEDs to know precisely the light emission start. Several flashes are done with the OB and the time residual between the OB emission and the PMT is measured for each PMT on the 7 storeys above the OB excluding the first one which is too close to the OB. The peak time of the time residual distribution is compared with the time expected from the light propagation and the calibrations done before the deployments. After the adjustments are done if it is needed.

An independent cross check of the inter calibration between the OMs in the same storey are done with the Čerenkov light from a potassium decay. There are about 150 photons from each decay which can be simultaneously seen in both OMs. If one plots the time difference between all hits for the pair of OMs a Gaussian peak of 10 ns FWHM from the ^{40}K emerges above the random background as it is seen in Fig. 2.17. Its mean should be at 0 if the time calibration is perfect.

There are two laser beacons which are used for the calibration of the time offsets between the lines. The laser beacons are powerful light sources which are placed on the bottom of two lines. The detected light on the other strings is compared with the estimated one from the line positions and the calibrations in the laboratory.

These time calibrations are performed regularly by the detector operators and then the calibration sets are created by the calibration experts. The current calibration scheme attains the relative time calibration between the detector elements of less than 1 ns.

2.3.2 Charge Calibration

Each ARS produces an integrated charge as a number of amplitude to voltage converter (AVC) channels. This is needed to be converted to a number of photoelectrons. For this a charge corresponding to one photon is needed to be measured together with the charge measured in the absence of the signal (pedestal) (Baret 2009). Both measurements are done regularly in-situ with the optical background which is mostly a

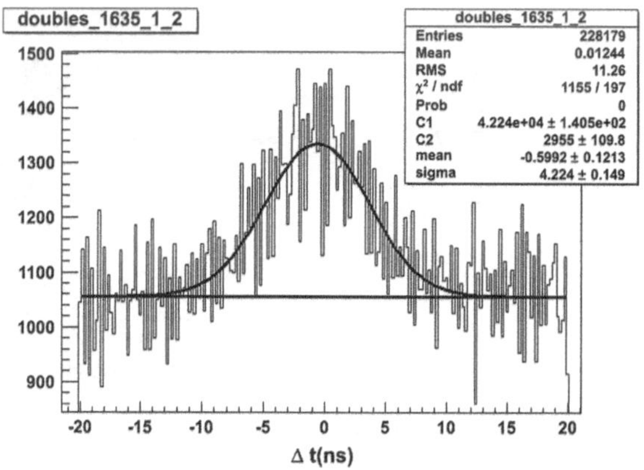

Fig. 2.17 An example of distribution of the time residual between hits on two PMTs

single photon signal. A 1 p.e. charge distribution is measured by writing the charge of all hits with a 0.3 p.e. threshold. A 0 p.e. charge distribution is obtained by integrating the charge at some random start time. Both obtained distributions are fitted with a Gaussian function and mean values are extracted. Using these values a charge in photoelectrons for any hit can be calculated as:

$$Q = \frac{AVC - AVC_{0\text{p.e.}}}{AVC_{1\text{p.e.}} - AVC_{0\text{p.e.}}} \tag{2.3}$$

where AVC is the measured charge in AVC channels, $AVC_{0\text{p.e.}}$ and $AVC_{1\text{p.e.}}$ are obtained 0 and 1 p.e. means from the calibrations.

If the pulse has an amplitude just above the threshold its time cannot be identified correctly by ARS and a zero value is produced. This hits do not affect the detector operation but may be used for the threshold tuning. Selecting time 0 hits it is possible to know the charge of the hits with an amplitude of the threshold set. The threshold can be tuned then for each ARS to 1/3 p.e. The values of effective thresholds, 0 and 1 p.e. changes with time and especially during the high voltage tunings which are done to have similar charge output for each PMTs. The above mentioned PMT parameters are monitored regularly and stored in the DataBase.

2.3.3 Position Calibration

The positions of the line anchors on the sea bed are well known and stable in time. The absolute positioning of each anchored detector component is calculated with an accuracy of about 1 m by acoustic triangulation from the surface with a ship equipped with a differential GPS. However the lines move freely in the water due to the current

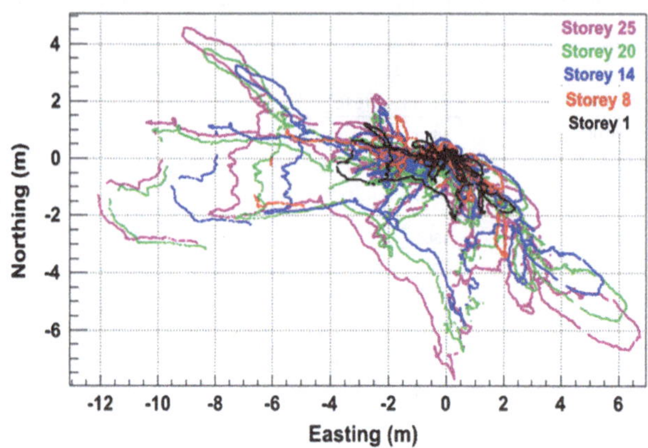

Fig. 2.18 The horizontal displacement of the hydrophones on one of the lines with respect to the anchor for a 6 months period

and the photomultipliers, especially on the top of the detector, change their absolute position and orientation. Their position is monitored by the system of the acoustic transponders and receivers distributed over the lines and on the sea bed (Adrián-Martínez et al. 2012). There are transmitters installed at the anchor of each line. Acoustic detectors—hydrophones are mounted on the storeys 1, 8, 14, 20 and 25 to detect signals in 40–60 kHz range. The distances are obtained by the triangulation of each receiver with respect to the emitters. This procedure is done regularly and the data is stored in the database. A relative movement of one of the detector lines is shown in Fig. 2.18. The LCM orientations are measured by compasses and tiltmeters in each storey. The current positioning strategy allows to monitor relative positions of all of the optical modules with an accuracy better than 20 cm.

References

S. Adrián-Martínez et al., The positioning system of the ANTARES Neutrino Telescope. JINST **7**(8), 1 (2012). T08002, ISSN 1748–0221, http://iopscience.iop.org/1748-0221/7/08/T08002/

M. Ageron et al., The ANTARES optical beacon system. NIM A **578**(3), 498–509 (2007). ISSN 01689002, http://linkinghub.elsevier.com/retrieve/pii/S0168900207012041

M. Ageron et al., ANTARES: the first undersea neutrino telescope. NIM A **656**(1), 11–38 (2011). ISSN 01689002, http://linkinghub.elsevier.com/retrieve/pii/S0168900211013994

J. Aguilar et al., Study of large hemispherical photomultiplier tubes for the ANTARES neutrino telescope. NIM A **555**(1–2), 132–141 (2005). ISSN 01689002, http://linkinghub.elsevier.com/retrieve/pii/S016890020501836X

J. Aguilar et al., The data acquisition system for the ANTARES neutrino telescope. NIM A **570**(1), 107–116 (2007). ISSN 01689002, http://linkinghub.elsevier.com/retrieve/pii/S0168900206017608

J. Aguilar et al., Performance of the front-end electronics of the ANTARES neutrino telescope. NIM A **622**(1), 59–73 (2010). ISSN 01689002, http://linkinghub.elsevier.com/retrieve/pii/S0168900210014154

J. Aguilar et al., AMADEUS—the acoustic neutrino detection test system of the ANTARES deep-sea neutrino telescope. NIM A, 626–627 (2011a). 128–143, ISSN 01689002, http://linkinghub.elsevier.com/retrieve/pii/S0168900210020772

J. Aguilar et al., Time calibration of the ANTARES neutrino telescope. Astropart. Phys. **34**(7), 539–549 (2011b). ISSN 09276505, http://linkinghub.elsevier.com/retrieve/pii/S0927650510002367

S. Aiello et al., The measurement of late and after-pulses in the large area Hamamatsu R7081 photomultiplier with improved quantum-efficiency photocathode, Technical report, INFN/TC-04/11 (2011)

P. Amram et al., The ANTARES optical module. NIM A **484**(1–3), 369–383 (2002). ISSN 01689002, http://linkinghub.elsevier.com/retrieve/pii/S0168900201020265

B. Baret, Charge calibration of the ANTARES high energy neutrino telescope, in Proceedings of the 31st ICRC, (ódź, 2009) http://icrc2009.uni.lodz.pl/proc/pdf/icrc1184.pdf

F. Feinstein, The analogue ring sampler: a front-end chip for ANTARES. NIM A **504**(1–3), 258–261 (2003). ISSN 01689002, http://linkinghub.elsevier.com/retrieve/pii/S0168900203007733

R. Gurin, A. Maslennikov, Controlhost: distributed data handling package, Technical report (1995) http://www.nikhef.nl/pub/experiments/antares/software/Dispatcher/ControlHost.ps

M. de Jong, The ANTARES Trigger Software, Technical report ANTARES-SOFT-2005-005 (2005)
B. Lubsandorzhiev et al., Photoelectron backscattering in vacuum phototubes. NIM A **567**, 12–16 (2006). ISSN 01689002, http://linkinghub.elsevier.com/retrieve/pii/S0168900206008606
D. Palioselitis, Measurement of the atmospheric neutrino energy spectrum, Ph.D. thesis, (Institute for High Energy Physics (IHEF), 2012) http://dare.uva.nl/cgi/arno/show.cgi?fid=444188

Chapter 3
Simulation of the ANTARES Detector Work

The detector response to the passage of particles is needed for the direction and energy reconstruction code development and testing. Additionally, all the background noises have to be simulated in order to estimate their impact on the detector performance. The simulation software package was developed, initially, to optimise the detector configuration during research and development phase. Lately, this software was improved, thanks to real data from the detector. The software consists of several parts: event generators (cosmic and atmospheric neutrinos, atmospheric muons), a light emission code, a signal digitisation and optical background simulation code.

Atmospheric muons are simulated with MUPAGE (Carminati et al. 2008) which is in a good agreement with the measurements done with the ANTARES detector itself (Aguilar et al. 2010). The neutrino events were simulated with GENHEN (Bailey 2002) which uses CTEQ6D (Pumplin et al. 2002) parton density functions for the deep inelastic charged current neutrino scattering cross-section calculation. This simulation is done at once for atmospheric and signal neutrinos with two separate weights. The weight for the atmospheric neutrinos is calculated using the Bartol group model (Agrawal et al. 1996). Cross sections and neutrinos propagation through the Earth were included into the weights calculation in both cases. The muons from the neutrino interactions were propagated till the detector volume with the MUSIC code (Antonioli et al. 1997) and hadronic shower production at the vertex inside the detector sensitive volume is simulated using GEANT[1] (Brunner 2003).

The light production from muons and showers, its propagation and detection is done with the sampling from the parametrised tables which include scattering and absorption in water measured at the ANTARES site (Aguilar et al. 2005). The optical module is parametrised as a flat area which detection efficiency is computed using proper GEANT simulations (Sect. 3.3.3). The simulation of an optical module output with its transit time spread function and afterpulses contributions is done according to the measurements. The environmental conditions of the detector for each data

[1] http://wwwasd.web.cern.ch/wwwasd/geant/.

V. Kulikovskiy, *Neutrino Astrophysics with the ANTARES Telescope*,
Springer Theses, DOI 10.1007/978-3-319-20412-3_3

taking run which is about 2–3 h are simulated. The so-called run by run simulations produce a simulated run for each data run. The detector calibration, alignment and the optical noise background are taken from the data for each run to use in simulation.

3.1 Atmospheric Background Simulation

Available predictions on the atmospheric neutrino fluxes show differences both in the absolute normalisation and in the energy-angular profiles. The model from the Bartol group (Agrawal et al. 1996) is used in this work and it is compared with FLUKA (Battistoni et al. 2003a, b) and Honda (1995) in the low energy range in Fig. 3.1 (this comparison is also published in Franco et al. 2013). Recent model updates include better propagation codes of the atmospheric showers (3D calculations), however, the largest uncertainties come from the primary cosmic-ray spectrum and the cosmic-ray interactions and here there is no significant improvements yet. As a result, these updated models did not modify the respective flux profile and amplitude predictions (Honda et al. 2011).

Several calculations of the fluxes from the charmed particles decay (the so-called prompt component) are shown in Fig. 3.2 in comparison with the conventional flux from the pions and kaons. The lack of precise information on high-energy charm production in hadron-nucleus collisions leads to the uncertainties of an order of magnitude for these models (Costa 2001; Enberg et al. 2008; Martin et al. 2003). The recent measurements with neutrino telescopes can not confirm the presence of the prompt component in the atmospheric neutrino flux due to the low events number at the energies above several TeV (Fusco and Kulikovskiy 2013; Schukraft 2013).

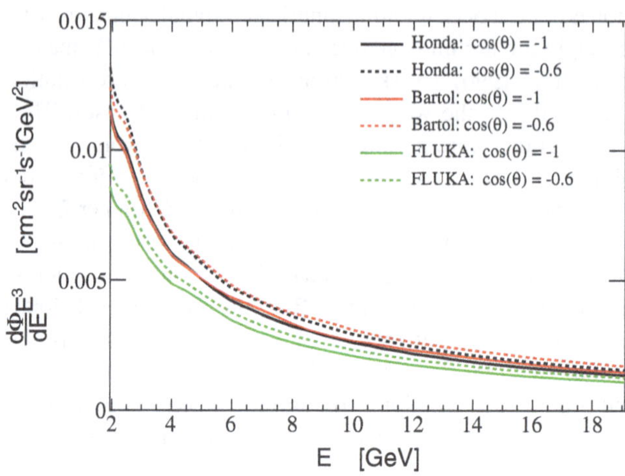

Fig. 3.1 Energy dependance of the atmospheric neutrino flux for three different models: Bartol (*red*), FLUKA (*green*) and Honda (*black*). *Solid* and *dashed lines* correspond to the fluxes incident vertically and upgoing at $\cos\theta = -0.6$) respectively

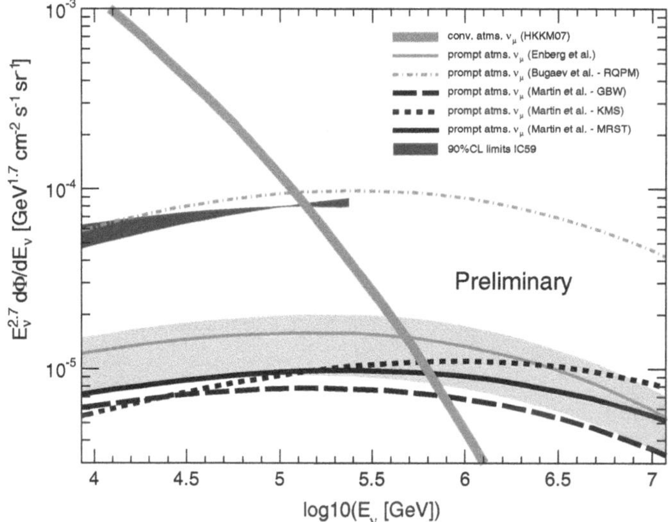

Fig. 3.2 Predictions for $\nu_\mu + \overline{\nu}_\mu$ prompt atmospheric neutrino fluxes in comparison to the expected flux of conventional atmospheric neutrinos. The *light grey shaped area* marks the theoretical uncertainty on the prediction by Enberg et al. (2008). The *dark grey shaded area* represents the limits on the atmospheric neutrino flux, obtained with the analysis of 348 days of the IceCube data in its 59-string configuration. The picture is taken from Schukraft (2013)

For the atmospheric muon simulation the MUPAGE software with a parametrised muon flux is used (Carminati et al. 2008). The parametrisation is obtained by the means of full Monte Carlo simulation software which was extensively tested with the MACRO underground experiment with a great overburden (Ambrosio et al. 1997). The input primary cosmic-ray flux is also taken from the MACRO model which is bounded by the measurements of the atmospheric muons with this experiment. The muons which originate from the decay of charmed hadrons are not included to the full simulation.

MUPAGE is capable to simulate bundles of muons from the single shower. This is a very important background component as the muons in the bundle are highly correlated in time and direction which may affect track and energy reconstruction algorithms.

3.2 Neutrino Event Generation

The instrumented detector volume is treated as a cylinder which contains all the PMTs. This volume is surrounded by the cylinder of a larger size which is called "can". Čerenkov light is generated only for the particles inside the can. The can size exceeds the instrumented volume by ~200 m which corresponds to at least three

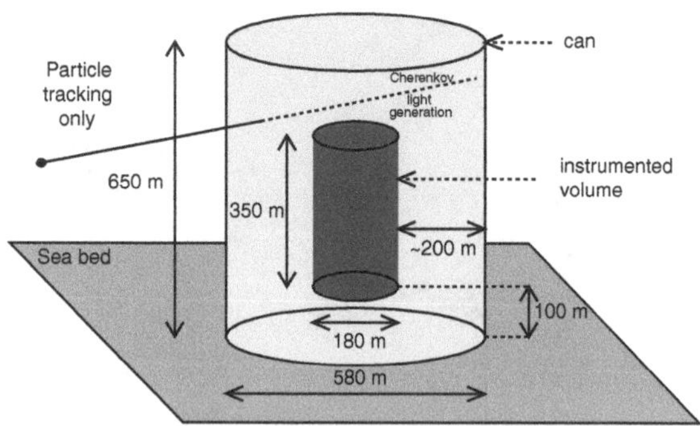

Fig. 3.3 Definition of the detector geometry for the event generation. The picture is taken from Bailey (2002)

light absorption lengths in the water. Outside only the particle energy losses are considered. The geometry of this simulation scheme is shown in Fig. 3.3.

The largest possible muon energy in simulation corresponds to the upper limit on the neutrino energies specified by a user, $E_{\rm max}$ with the value $E_{\rm max} = 10^8$ GeV is chosen for this work. From the muon energy the maximum muon ranges in both the rock and the water at this energy, $R_{\rm max}^{\rm rock}$ and $R_{\rm max}^{\rm water}$ respectively are calculated using the muon propagation code. By knowing these ranges, the total simulation volume is defined as a cylinder with a radius $R_{\rm max}^{\rm water}$, the top surface is at the height $R_{\rm max}^{\rm water}$ from the top of the instrumented volume and the bottom surface is at $R_{\rm max}^{\rm rock}$ from the sea floor.

Neutrinos are generated with a E^{-1} spectrum which can be reweighted with a cosmic $\sim E^{-2}$ spectrum or the atmospheric neutrino spectrum. The total interacting neutrino spectrum between $E_{\rm min}$ and $E_{\rm max}$ is divided into equal bins in $\log_{10} E_\nu$ (10 bins are used in the current work). For the ith bin with energies ($E^i_{\rm min}$,$E^i_{\rm max}$) the number of generated events is:

$$N^i(E^i_{\rm min}, E^i_{\rm max}) = N_{\rm total} \times \frac{\int_{E^i_{\rm min}}^{E^i_{\rm max}} E^{-1}\,{\rm d}E}{\int_{E_{\rm min}}^{E_{\rm max}} E^{-1}\,{\rm d}E}. \tag{3.1}$$

As the generated spectrum is quite different from the physical spectrums, the low energy neutrino events production in the range of 5–3000 GeV and the high energy neutrino production in the range of 10^2–10^8 GeV are done separately to ensure an high statistics of events in each energy bin.

For each energy bin a maximum muon free path in the water and the rock is recalculated and the "scaled" generation volume is calculated in the same way as it is done for the total simulation volume. The number of events to generate in that scaled volume is:

$$N^i_{\rm scaled}(E^i_{\rm min}, E^i_{\rm max}) = {\rm Poisson}\left(N^i \times \frac{V_{\rm scaled}}{V_{\rm total}}\right), \tag{3.2}$$

where the number of events to simulate was smeared with a Poisson distribution in order to the preserve the statistical fluctuations between different energy bins.

After for each event:

1. The neutrino energy is sampled from the E^{-1} spectrum.
2. Neutrino position is chosen within the scaled volume. The distance downwards the rock is scaled by the density to ensure that there are more interactions per unit volume in the rock compared to the water.
3. For events generated outside the can the shortest distance from the neutrino vertex position to the can is calculated. If the distance is greater than the maximum muon range at the generated neutrino energy, the event is discarded as no muon will ever reach the can for this event.
4. Neutrino direction is generated from an isotropic direction. For events outside the can the distance of the closest approach of the neutrino direction to the can is calculated and only events closer than the user defined cut distance (500 m in this work) are processed further. At the high neutrino energies the muon and neutrino tracks are closely correlated so this cut insures that the muon will reach the detector. At the lower energies the muon direction may appear quite different from the neutrino one, however the chosen cut distance is greater than the muon range so this condition becomes less stringent than the previous one.
5. For each event the interaction channel is chosen according to the relative cross-sections and the neutrino interaction is simulated using LEPTO (Ingelman et al. 1997) for deep inelastic scattering or RSQ (Barr 1987) for resonant or quasi-elastic scattering channels to get the final state particles at the neutrino interaction vertex. For deep inelastic scattering cross-sections CTEQ6D (Pumplin et al. 2002) parton density functions are used.
6. All particles are tracked inside the can. Tracking assumes the energy losses calculation, interactions and light emission. For the events outside the can only muons with tracks intersecting the can are propagated till the can surface and then tracked.
7. For events which are kept the event weight is calculated.

Neutrino event weight calculation includes the probability of neutrino absorption in the Earth. This probability depends on the density of matter along its path and the neutrino interactions cross-section. To good approximation, the Earth may be regarded as a sphere with a complex internal structure consisting of a dense inner and outer core and a lower mantle of medium density, covered by a transition zone, lid, crust, and oceans. Density profile of the Earth is taken from the Preliminary Earth Model (Gandhi et al. 1996).

For ν_e, ν_μ a charged-current interaction in the Earth far from the detector leads to the loss of the neutrino event. However, for the ν_τ the situation is more complicated since the produced tau decays on the flight with a production of ν_τ. This leads to the so-called regeneration chain thanks to which ν_τ is never absorbed. The energy losses due to the propagation through the Earth lead to a pile-up of tau neutrinos at the energies lower than the neutrino's energy at the Earth surface. Moreover, a tau can decay via muonic channel (BR $=$ 17.36 %) or via electronic channel (BR $=$ 17.84 %) with a production of secondary neutrinos. Neutral charged-current interaction cross-sections are lower than the charged-current cross-sections, however they can not be neglected, especially when searching for cascade-like events. These processes reduce the energy of the neutrinos passed through the Earth and increase the probability of this passage since the probability depends on the neutrino energy. All these processes are reproduced in GENHEN.

3.3 Optical Photons Simulation

The sea water is a uniformly transparent medium. The transparency of a medium may be characterised by the two main components—absorption and scattering. Each of them is a random process with an interaction length which depends from the wavelength. Also, there is a light produced by the radioactive elements decay and by the living organisms. This light presents optical background noise for the neutrino detection.

A vast amount of Čerenkov photons is needed to be simulated and propagated through the water. In the ANTARES simulation scheme this is realised by using fast approach where the number of detected photons is sampled from the so-called photon tables. For the description of the photons detection in the OMs a full GEANT simulation was performed and parametrised as angular acceptance.

3.3.1 Optical Properties of Water

Properties of the light absorption and scattering in the sea water close to the ANTARES site are estimated below. The estimation is based on theoretical models and different measurements, some of which were never published.

Scattering
Scattering may be expressed with volume-scattering function $\beta(\lambda, \theta)\,[\mathrm{m}^{-1}\mathrm{sr}^{-1}]$ which is the probability of a photon of wavelength λ being scattered by angle θ per unit length. This function can be rewritten using two other quantities: total scattering probability and the angular scattering functions. The first one can be expressed as following:

$$b(\lambda)[\mathrm{m}^{-1}] = \int_0^{\pi} 2\pi \sin(\theta)\beta(\lambda, \theta)\mathrm{d}\theta, \qquad (3.3)$$

while the angular scattering function is defined as:

$$\hat{\beta}(\lambda, \theta) = \frac{\beta(\lambda, \theta)}{b(\lambda)}, \qquad (3.4)$$

which are useful for scattering simulation. It is simple to check that the angular scattering probability integrated over the full angle gives unity:

$$\int_0^{\pi} 2\pi \sin(\theta)\hat{\beta}(\lambda, \theta)d\theta = 1. \tag{3.5}$$

For several scattering processes the total volume-scattering function can be written as:

$$\beta(\lambda, \theta) = \sum_{i=1}^{N_p} b_i(\lambda)\hat{\beta}_i(\lambda, \theta). \tag{3.6}$$

The theory of scattering in a pure water is derived by Smoluchowski (1908) and Einstein (1910). In this model the scattering on density fluctuations creates local refractive index fluctuations. The angular distribution either pure water or for pure sea water is wavelength independent and almost symmetric around 90°:

$$\hat{\beta}_{\mathrm{SE}}(\theta) = \left(\frac{3}{8\pi}\frac{1+\delta}{2+\delta}\right)\left(1 + \frac{1-\delta}{1+\delta}\cos^2\theta\right), \tag{3.7}$$

where $\delta = 0.09$ is the "depolarisation ratio". This coefficient is attributable to the anisotropy of the water molecules and it can be considered independent from salinity, pressure and temperature. The total scattering probability depends from wavelength in the following way:

$$b_{\mathrm{SE}}(\lambda) = b^0_{\mathrm{SE}}\left(\frac{550\ \mathrm{nm}}{\lambda}\right)^{4.32}, \tag{3.8}$$

where the power 4.32 is different from the "classic" Rayleigh scattering coefficient 4 due to the wavelength dependence of refraction index (Mobley 1994). In order to define the scattering of the pure water, the δ and b^0_{SE} were used from the measurements which can be done at the large scattering angles for the simplicity. The coefficient b^0_{SE} is dependent from the salinity according to the theory (the pure sea water values are about 30 % greater than the pure water values). Also it decreases for decreasing temperature or increasing pressure as both factors reduce the small-scale fluctuations. The measured value of $b^0_{\mathrm{SE}} = 1.94 \times 10^{-3}\ \mathrm{m}^{-1}$ can be found in Mobley (1994), Morel (1974), while $b^0_{\mathrm{SE}} = 1.7 \times 10^{-3}\ \mathrm{m}^{-1}$ value can be found in model (Kopelevich 1983) which is used in the ANTARES detector simulation. The salinity, pressure and temperature have variations lower than uncertainty of the pure water scattering at the ANTARES site.

The scattering on the particles which are contained in the natural sea water has a bigger impact than the scattering on density fluctuations. Figure 3.4 shows the measurements of three different types of sea water (Petzold 1972). Although the volume scattering distributions vary by a factor of 50, the angular scattering functions of

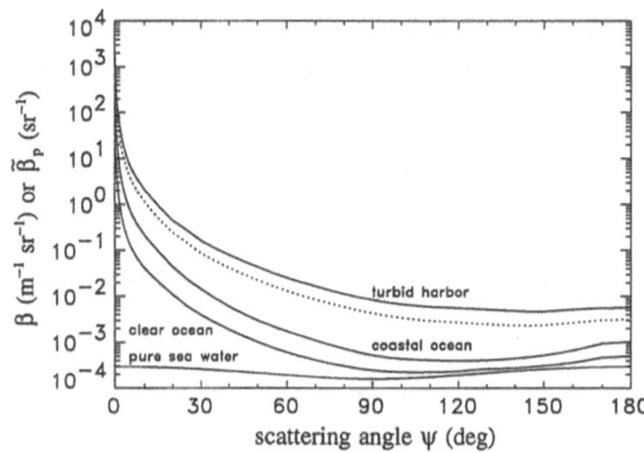

Fig. 3.4 Measured volume scattering functions (*solid lines*) from the three different natural water types, and the computed volume scattering function for the pure sea water, all at $\lambda = 514$ nm. The *dotted line* is the average angular scattering function $\hat{\beta}_{\mathrm{P}}(\theta, \lambda)$. This figure is taken from Mobley (1994, Fig. 3.13)

particle scattering are equal within measurement uncertainties. The average angular scattering function $\hat{\beta}_{\mathrm{P}}(\theta, \lambda)$ of the particle scattering was calculated and presented in Fig. 3.4 as dotted line. The theoretical description exists for the scattering on the spherical particles of the known size (the so-called "Mie scattering"). However, the natural sea water contains broad range of particle sizes from the tiny colloids and viruses of $0.1\,\mu$m size and about $100\,\mu$m from microplankton. The model of Kopelevich (1983) makes a simplification assuming two classes of particles: "large" and "small". The resulting volume scattering probability depends on their concentration:

$$\beta_{\mathrm{P}}(\theta, \lambda) = 1.34\nu_s\hat{\beta}_s(\theta)\left(\frac{550\ \mathrm{nm}}{\lambda}\right)^{1.7} + 0.312\nu_l\hat{\beta}_l(\theta)\left(\frac{550\ \mathrm{nm}}{\lambda}\right)^{0.3}, \quad (3.9)$$

where concentrations ν_s and ν_l are taken from the fit of the data in parts-per-million, $\beta_s(\theta)$ and $\beta_l(\theta)$ are tabulated in Kopelevich (1983). This model can be tuned to describe the total probability quite well. The systematic error of the model is claimed to be 30 % at maximum. Qualitatively, the angular scattering function (3.9) describes correctly the main trend of the measurements. However it lacks of precision. In the ANTARES model the average angular scattering function from the Petzold's measurements is used instead for $\beta_s(\theta)$ and $\beta_l(\theta)$. Also, the constants $\nu_s = \nu_l = 0.0075$ are used in the ANTARES water model. The scattering model which uses these constants fits quite well the measurements performed at the ANTARES site (Aguilar et al. 2005) although the values are slightly lower than the suggested ones ($0.01 \leq \nu_s \leq 0.20$ and $0.01 \leq \nu_l \leq 0.40$ ppm) in Kopelevich (1983). It is unclear, however, if this range is applicable to deep water. The stated accuracy of Kopelevichs model is of order 30 %, which also encompasses the difference between the ANTARES values of ν_s and ν_l hence, their low values should not be a concern.

Summarising, the volume scattering probability is described in ANTARES as following:

$$\beta(\theta,\lambda) = 1.7\times 10^{-3}\left(\frac{550\text{ nm}}{\lambda}\right)^{4.3}\hat{\beta}_{\text{SE}}(\theta)+ \\ +7.5\times 10^{-3}\hat{\beta}_{\text{P}}(\theta)\left(1.34\left(\frac{550\text{ nm}}{\lambda}\right)^{1.7}+0.312\left(\frac{550\text{ nm}}{\lambda}\right)^{0.3}\right). \tag{3.10}$$

Absorption

The absorption length measurements used preliminary in the ANTARES water model were performed in the Mediterranean Sea close to Greece (Khanaev and Kuleshov 1992). A highly collimated beam was used to measure an attenuation length which is a product of scattering and absorption. Attenuation probability is simply a sum of the absorption and scattering probabilities:

$$b_{\text{att}} = b_{\text{abs}} + b_{\text{scat}}, \tag{3.11}$$

while for the lengths which are inverse to the probabilities one can simply evaluate:

$$1/l_{\text{att}} = 1/l_{\text{abs}} + 1/l_{\text{scat}}. \tag{3.12}$$

The measurements of the attenuation length are done in the wavelengths range 300–600 nm with a step less than 20 nm. To obtain the absorption length the scattering length was subtracted from the measured attenuation length (Price 1997). For the scattering the following total probability was assumed:

$$b_{\text{scat}}(\lambda) = b^{0}_{\text{SE}}\left(\frac{550\text{ nm}}{\lambda}\right)^{4.32} + A\left(\frac{400\text{ nm}}{\lambda}\right)^{1.7}, \tag{3.13}$$

where $b^{0}_{\text{SE}} = 1.94\times 10^{-3}\ \text{m}^{-1}$ was used from Mobley (1994), Morel (1974). A was evaluated from the absorption length measurements with the uncollimated photostrobe. This photostrobe was used with a 460 nm interference filter at depths greater than 3500 m in the same location (Anassontzis and Ioannou 1994). Measured $l_{\text{abs}}(460\text{ nm}) = 55\pm 10$ m allowed to evaluate $A = 0.016\pm 0.004\ \text{m}^{-1}$.

In the ANTARES simulation the previously described measurements of absorption length first were corrected by using the measurements of absorption in the pure sea water (Smith and Baker 1981). The absorption length of the pure sea water was used whether the absorption length of the Mediterranean sea water appears to be longer (Fig. 3.5). Finally, the absorption length distribution was shifted on the wave length axis (Fig. 3.5) to fit the measurements with a AC9 transmissometer (Capone et al. 2002) performed at the ANTARES site (Capone 2004). AC9 transmissometer performs measurements with a collimated beam and the beam in a cylindrical mirror

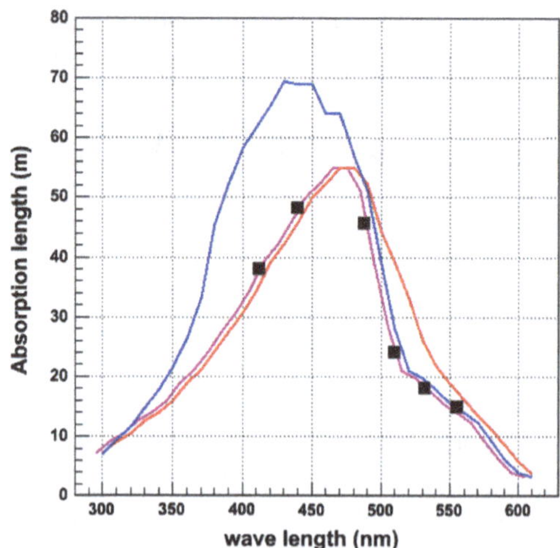

Fig. 3.5 Absorption lengths: the measurement of the Mediterranean sea water (*red line*) (Price 1997), the measurement of the pure sea water in the Sargasso sea (*blue line*) (Smith and Baker 1981), the measurement with the AC9 transmissometer (*black squares*) (Capone et al. 2002) performed at the ANTARES site, the absorption length fit, used in the ANTARES water model (*purple line*)

tube at different wavelengths allowing to measure absorption and attenuation lengths. The obtained absorption length curve is in agreement with measurements at the ANTARES site with a blue and UV light-emitting diodes (LEDs) (Aguilar et al. 2005).

3.3.2 Implementation of the Optical Photons Simulation

A full simulation of optical photons in which every Čerenkov photon is generated and propagated individually is not possible due to the very large number of photons produced by single muon passing through the detector and the need for high statistics of muons, especially in the study of the atmospheric background. Therefore, an alternative simulation of the photon scattering is needed and in the ANTARES simulation this is achieved by building "photon tables" which store the distributions of the numbers and arrival times of the PMT hits at different distances, positions and orientations with respect to a given muon track or an electromagnetic shower. The KM3 package of programs provides the simulation of the ANTARES detector response to the passage of high energy muons including the effect of the photon scattering in the water during the photon tables creation (Navas and Thompson 1999).

The first stage in the generation of the photon tables is to perform simulation of either short sections of muon tracks or electromagnetic showers and to record the distributions of the photons arriving on series of spherical surfaces surrounding the simulated particles. GEN is a program which simulates the generation of Čerenkov light by a particle in the water, including light from any secondary particles. A complete GEANT simulation is used at this step and the scattering is taken into account, recording the position, direction and arrival time of photons at spherical shells of various radii centred on the origin. This process is schematically shown in Fig. 3.6.

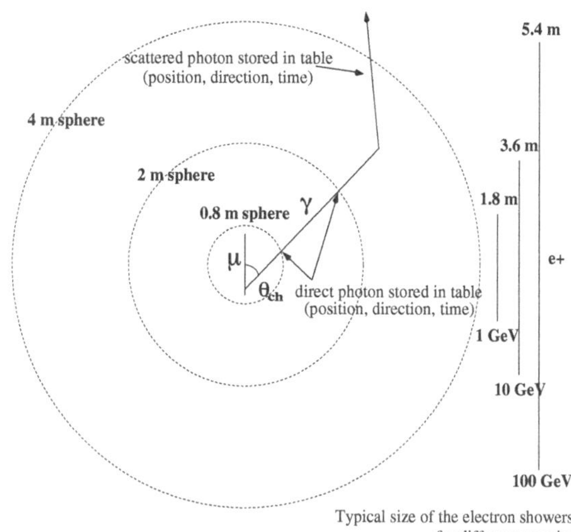

Fig. 3.6 Muons and electrons are propagated through the medium and Čerenkov photons are stored in spheres at different distances. The picture is taken from Navas and Thompson (1999)

The output of GEN contains a complete record of the photon properties (position, direction, wavelength, time) as they cross each of the spherical shells above. At the final step (HIT) of the photon tables creation, the response of the ANTARES optical modules, is needed to be added in order to produce a number of detected photons. This is done by dividing each shell into several bins in $\cos\theta$ (referred to as "bands") and looping over different OM orientations, folding in the photons arriving in that band with the optical module efficiency as a function of wavelength and the angle between the photons and the OM orientation. This optical module efficiency dependence from the photons incident angle is taken from the full GEANT simulations of the optical module, described in Sect. 3.3.3. For each case (muon or showers of a given energy) four OM hit distributions are produced:

- Probability of a direct photon (no scattering).
- The inverse cumulative time distribution of direct photons.
- Probability of a scattered photon.
- The inverse cumulative time distribution of scattered photons.

The main detector Monte Carlo simulation program (KM3MC) takes the photon tables generated by GEN and HIT above and uses them to simulate the detector response to the passage of high energy muons. The muon simulation is performed by MUSIC in short steps and, for muons close to the minimum ionising energy losses, photons arrival time to a PMT is drawn from the muon table created above. For sections of muon tracks with an energy loss above some threshold an electromagnetic shower is defined and the number of photons is drawn from the electron tables.

3.3.3 OM Simulation

The probability to detect the photons arriving to the OM from different angles in the water, the so-called angular acceptance is needed to build the photon tables described above. The angular acceptance of the ANTARES OM has been measured by the ANTARES collaboration in Saclay and Genova. In both experiments the OM was positioned in a water tank and the angular acceptance was measured by detection of the Čerenkov light produced by down-going muons passing through the water tank. By rotating the OM it was possible to detect the photons at different angles between the axis of the Čerenkov front and the axis of the OM and therefore to obtain the angular acceptance. Both experiments suffered from the photon scattering in the water that affected the angular acceptance measurement especially at large angles where the efficiency of detecting photons is much lower. To understand better the measurement, a GEANT4 simulation was developed with my contribution to describe the ANTARES OM (Anghinolfi et al. 2009).

The shape and materials of the PMT and OM were taken from the technical drawings. For the detailed simulation the photocathode thickness and refractive index are fundamental parameters. For a bialkali photocathode these two parameters have been obtained by a dedicated measurement reported in Motta and Schönert (2005).

Once the optical photon is absorbed by the photocathode, the probability that the conversion to a photoelectron occurs, depends on the quantum efficiency (Fig. 3.7). The measurement of the quantum efficiency was provided by the PMT manufactory (Hamamatsu). The produced photoelectrons not always reach the first dynode and create a detectable electron cascade in the dynode stack. The detection of the photoelectrons is described by the so-called collection efficiency, which depends on the radial position on the photocathode of the impinging photon, due to the electric field configuration inside the PMT. The collection efficiency was not available by the direct measurements. It was decided to extract it from the OM angular acceptance measurements with laser in air comparing it with the simulation of the same process.

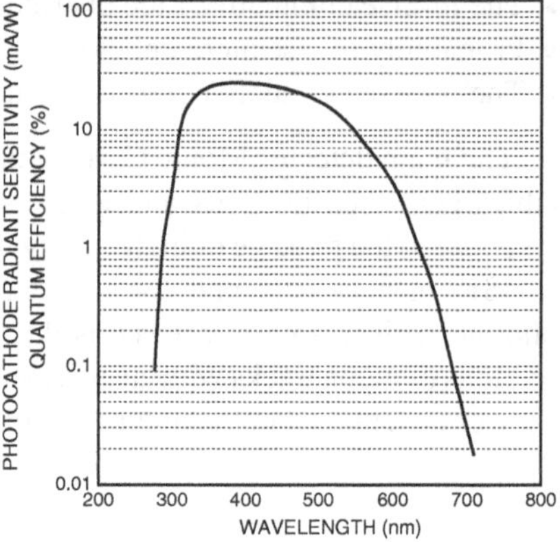

Fig. 3.7 Hamamatsu R7081-20 quantum efficiency. The image is taken from Hamamatsu (2008)

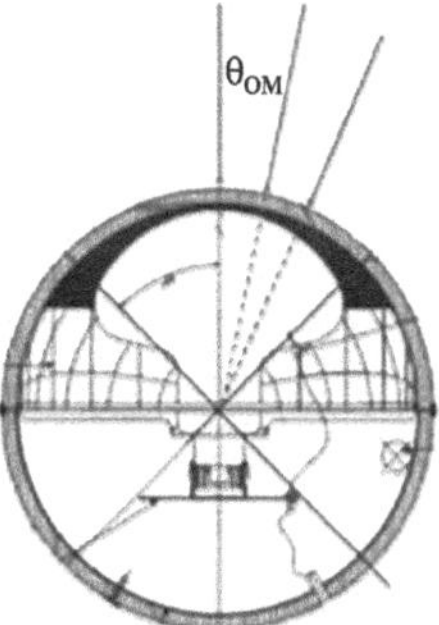

Fig. 3.8 Scheme of the radial scan measurement. The laser was always perpendicular to the OM surface

All simulation parameters except the collection efficiency are known so the last one was tuned to match the measurements with the laser.

Radial scans (Fig. 3.8) of two OMs were performed at Genova and at the University of Erlangen for different PMT azimuthal orientations. The experimental setup was a laser or a led impinging normally on the OM surface and the measurement consisted in the photon detection in dependence from the angle between the laser and the OM axis. Example of these scans is shown in Fig. 3.9. The local minimums at 0, 20 and 40° correspond to the photons falling on the μ-metal cage. The peak at 50° is due to reflection in the optical gel. A simulation of the setup with the laser was done using GEANT4 to tune the collection efficiency. The probability curve of the collection efficiency was manually developed using total radial acceptance curve as a starting point for the collection efficiency curve and changing it until the experimental radial acceptance curve could be reproduced in the simulation.

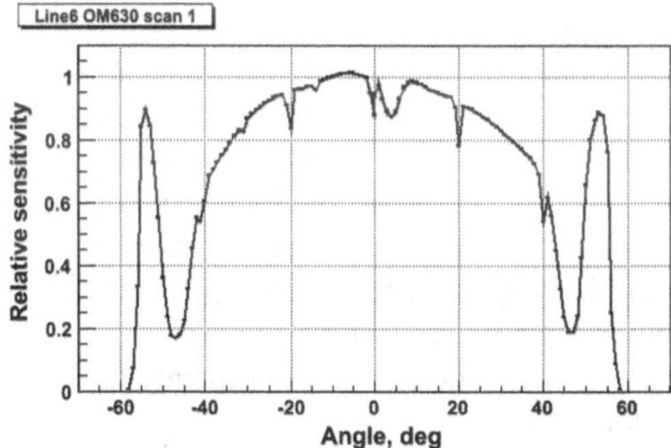

Fig. 3.9 Example of the measured relative detection efficiency of the photons arriving to OM. The measurement was done at the University of Erlangen using highly collimated laser beam arriving at different angle in respect to the PMT orientation (zero corresponds to the front side of the PMT). Scan is done at fixed azimuth angle. Some asymmetry is seen which may be ascribed due to the photocathode, OM glass non-uniformities and asymmetry of the dynode stack and the metal cage

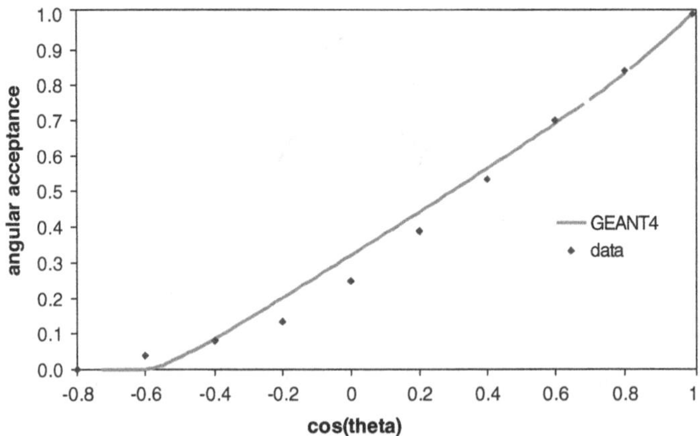

Fig. 3.10 Comparison between angular acceptance curves: GEANT4 simulations (Anghinolfi et al. 2009) and the measurements in the water tank (Amram et al. 2002)

The resulting angular acceptance obtained from the simulation of the photon flux in water without scattering and absorption is shown in Fig. 3.10 in comparison with the measurements in the water tank (Amram et al. 2002). It was used for the ANTARES detector simulation described in the next chapter. The OM simulation with GEANT4 was also used for the estimation of the signal coming from supernovae and the possibility of their detection (Kulikovskiy 2011).

3.3.4 Optical Background

The optical background of the deep sea water is described by two components: the radioactive elements decay and the light from living organisms (bioluminescence). The main radioactive component creating a background noise is ^{40}K present in salt. Decay rate of the radioactive elements is constant in time, depending only on the salinity. One decay usually provides several photons which can be seen simultaneously in different PMTs of the detector. The bioluminescence is defined as a light produced by living organisms. Single photons are released by a chemical reaction (chemiluminescence). Both optical background noises present the main source of the light in PMTs put in the deep sea water.

The radioactive decay background is present mainly due to the potassium from the dissolved salt in the water:

$$^{40}\mathrm{K} \rightarrow {}^{40}\mathrm{Ca} + e^- + \bar{\nu}_e\,(\mathrm{BR} = 80.3\,\%) \tag{3.14}$$

$$^{40}\mathrm{K} + e^- \rightarrow {}^{40}\mathrm{Ar} + \nu_e + \gamma\,(\mathrm{BR} = 10.7\,\%). \tag{3.15}$$

The decay energy of both reactions is above 1.3 MeV. This allows the Čerenkov emission for the high energetic tail of the beta-decay spectrum of electron. For the gamma rays, the Compton scattering is possible with a production of an electron above the Čerenkov threshold. The expected rate from the decay depends only on a salinity. The salinity variation with time and depth at the ANTARES site is shown in Fig. 3.11. For the ANTARES detector simulation this optical background is assumed to be constant. A single PMT rate was estimated with the detailed GEANT4 simulation on level of 20–40 kHz depending on a model choice for the light absorption in the sea water.

Bioluminescence is a form of chemiluminescence where light energy is released by a chemical reaction. Luciferin is oxidised with a molecular oxygen with the formation of a product molecule in an electronically excited state (Hastings 1996). De-excitation is followed by a photon emission. Luciferins (literally "light bearers") fall into many unrelated chemical classes. For example, a bacterial luciferin consists of a long-chain aldehyde and a reduced riboflavin phosphate. In the deep sea water the bioluminescence emission usually has a spectrum with a maximum in the range from 450 to 490 nm which is optimal for the light transmission (compare with Fig. 3.5). Additional shifters (luminophore molecules) and filters can be used by organisms to produce light emission of a desired wavelength.

The bioluminescence presents a varying background. An example of a PMT background rate is shown in Fig. 3.12. The peaks of bioluminescence of several hundreds kHz are seen. This rapidly changing background is though to originate from the macroscopic animals which are stimulated to emit bioluminescent light, for instance when suffering from the turbulent water or during the contact with the detector structure (Brunner 2011). Additionally, the mean background rate of each PMT is changing slowly. This may be explained by the overall change of concentration of the bioluminescent organisms in the sea water of the detector. In Fig. 3.13 the mean detector rate for one PMT is shown. Several high bioluminescence periods are seen, typically arriving in spring. These periods are connected with the water exchange from the surface layers enriched by the luminescent plankton (van Haren et al. 2011).

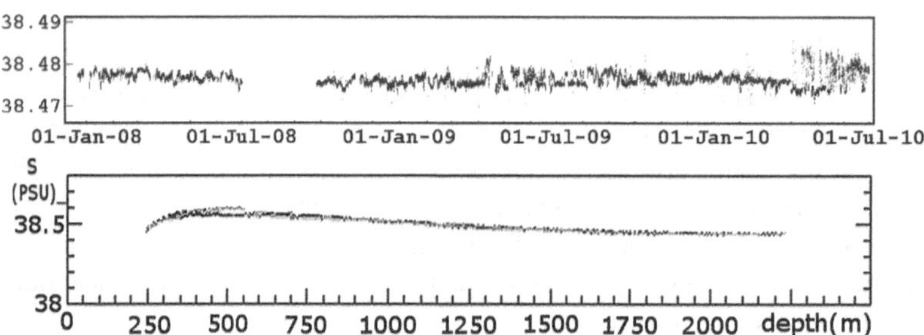

Fig. 3.11 Salinity variation with time (*top*) (Tamburini et al. 2013) and depth (*bottom*) (Capone 2004) at the ANTARES site

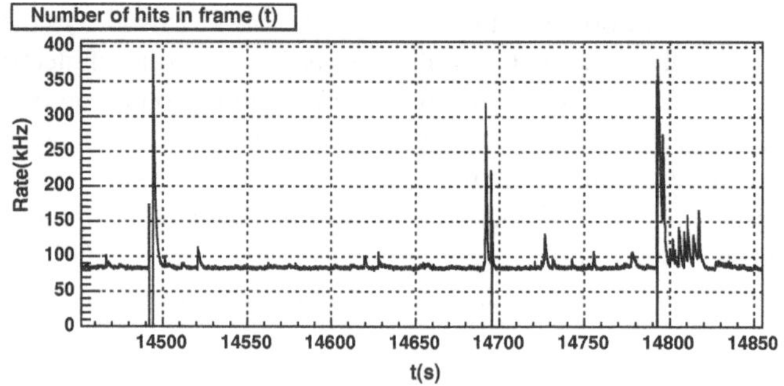

Fig. 3.12 An example of the PMT rate at the ANTARES site

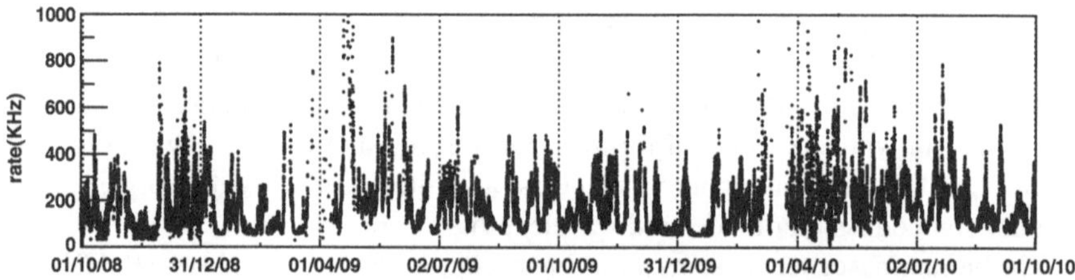

Fig. 3.13 Mean PMT rate in the ANTARES detector during two years of data taking

A strong water convection is ascribed to the dense water formation processes during winter and early spring. Additionally, the instabilities of the current close to the coast, in response to the atmospheric forcing above the deep open subbasin, creates mesoscale water motions. This phenomenon is assumed to create variations in the water exchange at the ANTARES site. High velocity water motion disturbs the luminescent macro-organisms (for instance when suffering from turbulent water or at contact with the detector structure) and they in turn produce the bioluminescent bursts. It was shown that the burst rate follows the current speed (Brunner 2011, Fig. 4). Figure 3.14 illustrates clearly this correlation. It can be seen that the bioluminescence follows absolute current velocity and not only one of its particular component. Also, average detector rate together with each PMT rate are changing in the same phase (better seen in Fig. 3.15). This means that light arrives from bioluminescent organisms around the PMTs rather than from organisms hitting PMTs directly. The periodicity of less than a day is seen in the optical rate and in the water velocity. This period corresponds to the inertial periodicity of the water motion.

The inertial period appears from the Euler equation written for the ideal liquid in the presence of the Earth rotation:

$$\frac{d\vec{v}}{dt} = -\frac{1}{\rho}\vec{\nabla}p - 2\vec{\Omega} \times \vec{v} - \vec{g}, \tag{3.16}$$

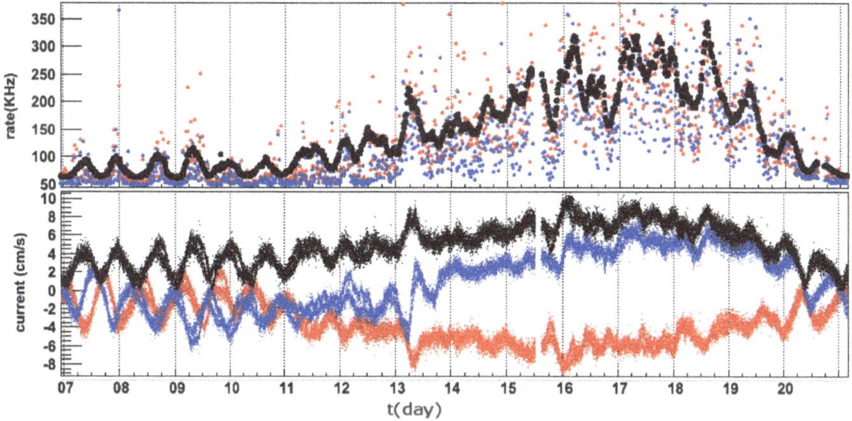

Fig. 3.14 Variation of PMT rates (*above*) in time: average over all PMTs (*black*), PMT 1 (*red*) and PMT 2 (*blue*) on a particular floor together with the absolute sea current velocity (*below*): total (*black*), north projection (*blue*), east projection (*red*)

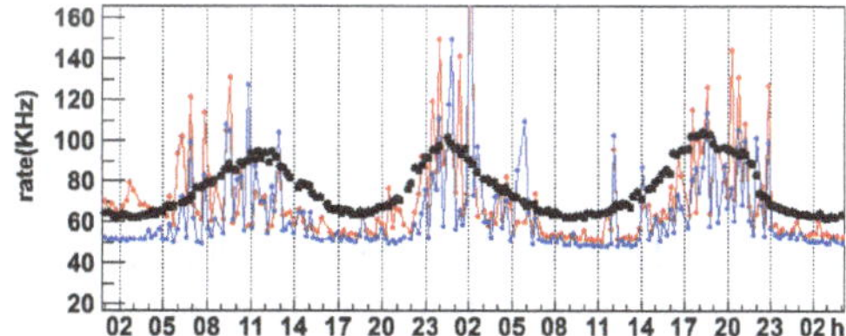

Fig. 3.15 PMT rates of the ANTARES detector: average over all PMTs (*black*), particular floor PMT 1 (*red*) and PMT 2 (*blue*) as a function of time (hour)

where $\vec{v} = (v, u, w)$ is the velocity vector, ρ is the water density, p is the pressure, $\vec{\Omega}$ is the Earth rotation vector and $\vec{g}$ is the acceleration due to gravity (Stewart 2008, Chap. 9). One can simplify this equation for projections, assuming there is no horizontal pressure component (which is true if there are no additional forces) and no vertical water motion (which is a good assumption for deep water volumes close to the flat sea bottom far from the shore) as following:

$$\begin{cases} \dfrac{du}{dt} &= 2\Omega v \sin\phi \\ \dfrac{dv}{dt} &= -2\Omega u \sin\phi \end{cases} \tag{3.17}$$

From this system of equation the simple harmonic oscillator equation can be derived for each of the horizontal components:

$$\frac{d^2u}{dt^2} + (2\Omega \sin\phi)^2 u = 0, \tag{3.18}$$

where ϕ is the latitude. The inertial frequency is:

$$\omega = 2\Omega \sin \phi, \tag{3.19}$$

which gives period for the ANTARES site:

$$T = 2\pi/(2\Omega \sin \phi) = \pi/(7.292 \times 10^{-5}\ [\text{rad/s}] \times \sin(43.1^\circ)) = 17.5\,\text{h}. \tag{3.20}$$

The solution of two coupled differential Eq. 3.17 is the velocity vector, which changes its direction without changing the absolute value:

$$\begin{aligned} u &= A \sin(\omega t) \\ v &= A \cos(\omega t) \end{aligned} \tag{3.21}$$

The rotation is clockwise in the Northern Hemisphere, counterclockwise in the Southern. The phase shift of $T/4 \approx 4.4$ h can be seen for the north and east current velocity projections in Fig. 3.14. In the presence of strong water convection and mesoscale water motions, the water motion can be described as following:

$$\begin{aligned} u &= B_u(t) + A \sin(\omega t) \\ v &= B_v(t) + A \cos(\omega t), \end{aligned} \tag{3.22}$$

where $B_u(t)$ and $B_v(t)$ are slowly changing functions. Their presence keeps $\sin(\omega t)$ and $\cos(\omega t)$ terms for the absolute current velocity. That is why oscillation patterns are seen in Fig. 3.14 for the absolute value of current velocity.

3.3.5 Optical Background Noise and PMT Electronic Response Simulation

Optical background from the ^{40}K decay and the bioluminescence should be added to simulation. In the so-called run by run scheme the optical background rates for each PMT are taken from the real data from each run. The optical background photons are simulated according to the extracted rates and added to the photons from the muon events. The simulated run is produced for each data run.

For each detected photon its time is smeared according to the PMT TTS (Fig. 2.11). TTS for each PMT was measured at Saclay during the OM integration and verified in Genoa during the PMT studies. Charge of each detected photon is drawn from the Gaussian distributions (similar to which is shown in Fig. 2.10) obtained during the charge calibration procedure. Additionally, small smearing of the charge calibration parameters (pedestal and 1 p.e. peak, Sect. 2.3.2) is done to introduce uncertainty during the charge calibration. Afterpulses according to the measurements done with real data are also added at this stage. For the known charge and time of each detected photon ARS digitisation procedure is simulated to introduce dead times (Sect. 2.2.1). Finally the events triggering is done on the simulated data. The trigger setup is taken accordingly to the real data runs. At the end the simulated data runs have the same properties of the optical background and similar detector configuration including detector calibration and trigger setup as the real data runs (de Jong 2010).

References

V. Agrawal et al., Atmospheric neutrino flux above 1 GeV. Phys. Rev. D **53**(3), 1314–1323 (1996). ISSN 0556–2821, http://www.ncbi.nlm.nih.gov/pubmed/10020123

J. Aguilar et al., Transmission of light in deep sea water at the site of the Antares neutrino telescope. Astropart. Phys. **23**(1), 131–155 (2005). ISSN 09276505, http://linkinghub.elsevier.com/retrieve/pii/S0927650504001902

J. Aguilar et al., Zenith distribution and flux of atmospheric muons measured with the 5-line ANTARES detector. Astropart. Phys. **34**(3), 179–184 (2010). ISSN 09276505, http://linkinghub.elsevier.com/retrieve/pii/S0927650510001246

M. Ambrosio et al., High energy cosmic ray physics with underground muons in MACRO. I. Analysis methods and experimental results. Phys. Rev. D **56**(3), 1407–1417 (1997). ISSN 0556–2821, http://link.aps.org/doi/10.1103/PhysRevD.56.1407

P. Amramet al., The ANTARES optical module. NIM A **484**(1–3),369–383 (2002). ISSN 01689002, http://linkinghub.elsevier.com/retrieve/pii/S0168900201020265

E. Anassontzis, P. Ioannou, Light transmissivity in the NESTOR site. NIM A **349**, 242–246 (1994). http://www.sciencedirect.com/science/article/pii/0168900294906270

P. Antonioli et al., A three-dimensional code for muon propagation through the rock: MUSIC. Astropart. Phys. **7**(4), 357–368 (1997). ISSN 09276505, http://linkinghub.elsevier.com/retrieve/pii/S0927650597000352

M. Anghinolfi et al., GEANT4 simulation of the ANTARES Optical Module, Technical report ANTARES-OPMO-2009-003 (2009)

G.D. Barr, The Separation of Signals and Background in a Nucleon Decay Experiment., Ph.D. thesis, University of Oxford (UK), 1987

D. Bailey, Monte Carlo tools and analysis methods for understanding the ANTARES experiment and predicting its sensitivity to Dark Matter, Ph.D. thesis, Wolfson College, Oxford, 2002, http://ethos.bl.uk/OrderDetails.do?uin=uk.bl.ethos.249189

G. Battistoni et al., Erratum to The FLUKA atmospheric neutrino flux calculation [Astropart. Phys. 19 (2003) 269 290]. Astropart. Phys. **19**(2), 291–294 (2003a). ISSN 09276505, http://linkinghub.elsevier.com/retrieve/pii/S0927650503001075

G. Battistoni et al., The FLUKA atmospheric neutrino flux calculation. Astropart. Phys. **19**(2), 269–290 (2003b). ISSN 09276505, http://linkinghub.elsevier.com/retrieve/pii/S0927650502002463

J. Brunner, Antares simulation tools, ed. by E. de Wolf. in *Proceedings of the workshop on Technical Aspects of a Very Large Volume Neutrino Telescope in the Mediterranean Sea*, (Amsterdam, The Netherlands, 2003), pp. 109–113, http://www.vlvnt.nl/proceedings.pdf

J. Brunner, The ANTARES neutrino telescope—Status and first results, NIM A, 626–627 (2011), S19–S24, ISSN 01689002, http://linkinghub.elsevier.com/retrieve/pii/S0168900210014221

A. Capone et al., Measurements of light transmission in deep sea with the AC9 trasmissometer, NIM A **487**(3), 423–434, (2002) ISSN 01689002, http://linkinghub.elsevier.com/retrieve/pii/S0168900201021945

A. Capone, Studio del sito di Capo Passero, in *NEMO collaboration meeting 2004*, (Genova, 2004) http://people.na.infn.it/barbarin/MaterialeScientifico/NEMO/Site/Capone_Genova04.ppt

G. Carminati et al., Atmospheric muons from parametric formulas: a fast generator for neutrino telescopes (MUPAGE), Comput. Phys. Commun. **179**(12), 915–923, (2008) ISSN 00104655, http://linkinghub.elsevier.com/retrieve/pii/S001046550800266X

C. Costa, The prompt lepton cookbook, Astropart. Phys **16**(2), 193–204, (2001) ISSN 09276505, http://linkinghub.elsevier.com/retrieve/pii/S0927650501001050

M. de Jong, The TriggerEfficiency program, Technical report ANTARES-SOFT-2009-001 (2010)

A. Einstein, Theorie der Opaleszenz von homogenen Flüssigkeiten und Flüssigkeitsgemischen in der Nühe des kritischen Zustandes, Ann. d. Phys. **338**(16), 1275–1298, (1910) http://onlinelibrary.wiley.com/doi/10.1002/andp.19103381612/abstract

R. Enberg et al., Prompt neutrino fluxes from atmospheric charm, Phys. Rev. D **78**(4), (2008), 043005, ISSN 1550–7998, http://link.aps.org/doi/10.1103/PhysRevD.78.043005

D. Franco et al., Mass hierarchy discrimination with atmospheric neutrinos in large volume ice/water Cherenkov detectors, JHEP (4), (2013), 008, ISSN 1029–8479, http://link.springer.com/10.1007/JHEP04(2013)008

L. Fusco, V. Kulikovskiy, Measurement of the atmospheric $\nu\_$ energy spectrum with the ANTARES neutrino telescope, in *Proceedings of the 33rd ICRC*, (Rio de Janeiro, Brasil, 2013) http://www.cbpf.br/icrc2013/papers/icrc2013-0636.pdf

R. Gandhi et al., Utrahigh-energy neutrino interactions, Astropart. Phys. **5**, 81–110, (1996) http://www.sciencedirect.com/science/article/pii/0927650596000084

J. Hastings, Chemistries and colors of bioluminescent reactions : a review, Gene **173**, 5–11, (1996), http://www.sciencedirect.com/science/article/pii/0378111995006761

Hamamatsu, Hamamatsu large photocathode area photomultiplier tubes, Technical report Hamamatsu, (2008), http://www.hamamatsu.com/resources/pdf/etd/LARGE_AREA_PMT_TPMH1286E05.pdf

M. Honda et al., Calculation of the flux of atmospheric neutrinos, Phys. Rev. D **52**(9), (1995), http://prd.aps.org/abstract/PRD/v52/i9/p4985_1

M. Honda et al., Improvement of low energy atmospheric neutrino flux calculation using the JAM nuclear interaction model, Phys. Rev. D **83**(12), 123001, (2011), ISSN 1550–7998, http://link.aps.org/doi/10.1103/PhysRevD.83.123001

G. Ingelman et al., LEPTO 6.5 A Monte Carlo generator for deep inelastic lepton-nucleon scattering, Comput. Phys. Commun. **101**(1–2), 108–134, (1997), ISSN 00104655, http://linkinghub.elsevier.com/retrieve/pii/S0010465596001579

S. Khanaev, A. Kuleshov, Measurements of water transparency South-West of Greece, in ed. by L. Resvanis, *Proceedings of the Second NESTOR International Workshop*, University of Athens, Greece, pp. 253–269, http://www.inp.demokritos.gr/web2/nestor/www/2nd/files/253_269_khanaev.pdf

O. Kopelevich, in *Small-parameter model of optical properties of sea water*, ed. by A. Monin, Ocean Optics, vol. 19, chap. 8, Nauka (1983)

V. Kulikovskiy, SN neutrino detection in the ANTARES neutrino telescope, in *Proceedings of the 32nd ICRC* (Beijing, China, 2011), http://arxiv.org/abs/1112.0478

A. Martin et al., Prompt Neutrinos from atmospheric cc and bb production and the Gluon at Very Small x, Acta Phys. Pol. B **34**, 3273–3303, (2003), http://www.actaphys.uj.edu.pl/_cur/store/vol34/pdf/v34p3273.pdf

C. Mobley, in *Light and Water: Radiative Transfer in Natural Waters*, (Academic Press, San Diago), cd edition ed. (1994) http://www.curtismobley.com/LightandWater.zip

A. Morel, Optical properties of pure water and pure sea water, in *Optical aspects of oceanography* (1974), http://www.obs-vlfr.fr/LOV/OMT/fichiers_PDF/Morel_OAO_74.pdf

D. Motta, S. Schönert, Optical properties of bialkali photocathodes, NIM A **539**(1–2), 217–235, (2005), ISSN 01689002, http://linkinghub.elsevier.com/retrieve/pii/S0168900204022132

S. Navas, L. Thompson, KM3 User Guide and Reference Manual, Technical report, ANTARES-SOFT-1999-011 (1999)

T. Petzold, in *Volume Scattering Functions for Selected Ocean Waters*. SIO Reference (Scripps Institution of Oceanography, 1972), pp. 72–78, http://oai.dtic.mil/oai/oai?verb=getRecord&metadataPrefix=html&identifier=AD0753474

P.B. Price, Implications of optical properties of ocean, lake, and ice for ultrahigh-energy neutrino detection. Appl. Optics **36**(9), 1965–1975, (1997), ISSN 0003–6935, http://www.ncbi.nlm.nih.gov/pubmed/18250887

J. Pumplin et al., New generation of parton distributions with uncertainties from global QCD analysis, JHEP 07, (2002), http://iopscience.iop.org/1126-6708/2002/07/012

A. Schukraft, A view of prompt atmospheric neutrinos with IceCube, Nucl. Phys. B - P. Sup. 237–238, 266–268, (2013), ISSN 09205632, http://linkinghub.elsevier.com/retrieve/pii/S0920563213002260

R.C. Smith, K.S. Baker, Optical properties of the clearest natural waters (200–800 nm), Appl. Optics **20**(2), 177–184, (1981), ISSN 0003–6935, http://www.ncbi.nlm.nih.gov/pubmed/20309088

M. Smoluchowski, Molekular-kinetische Theorie der Opaleszenz von Gasen im kritischen Zustande, sowie einiger verwandter Erscheinungen. Ann. d. Phys. **330**(2), 205–226, (1908), http://adsabs.harvard.edu/abs/1908AnP...330.205S

R.H. Stewart, Introduction to physical oceanography, Texas A & M University, September 2008, http://oceanworld.tamu.edu/home/course_book.htm

C. Tamburini et al., Deep-sea bioluminescence blooms after dense water formation at the ocean surface. PLOS ONE **8**(7), e67523, (2013), ISSN 1932–6203, http://www.pubmedcentral.nih.gov/articlerender.fcgi?artid=3707865&tool=pmcentrez&rendertype=abstract

H. van Haren et al., Acoustic and optical variations during rapid downward motion episodes in the deep north-western Mediterranean Sea, Deep-Sea Res. Pt. I **58**(8), 875–884, (2011), ISSN 09670637, http://linkinghub.elsevier.com/retrieve/pii/S0967063711001166

Chapter 4
Event Reconstruction and Data Selection

4.1 Event Reconstruction

The main goal of this section is the explanation of the muon event reconstruction which is used in this work. Muon event reconstruction is done in two steps: a track reconstruction and an energy reconstruction. The energy reconstruction algorithms usually require the knowledge of the track. It is important first to describe in more details the muon passage through the matter.

4.1.1 Track Reconstruction

The track of a muon passing through the detector is reconstructed using the arrival time of the photons together with the positions and the orientations of the photomultipliers. The track reconstruction used in this work is a multistep fitting algorithm (Heijboer 2002, 2004). Each step uses the result of the previous procedure as a starting point, with the exception of the first procedure, which is linear and therefore does not require a starting point. Some of these procedures are improved versions of algorithms that already existed in ANTARES.

The algorithm is designed for the reconstruction of muon tracks with energies above ~50 GeV. The majority of these muons will traverse the whole detector. Both angular deviation from a straight line (due to multiple scattering) and the deviation of the muon velocity from the velocity of light c (due to finite energy) are tiny. Therefore, through-going muons can be characterised by the position $\vec{p}$ of the muon at a fixed time t_0 and its direction $\vec{d}$. The direction can be parametrised in terms of the azimuth and zenith angles θ and ϕ: $\vec{d} = (\sin\theta\cos\phi, \sin\theta\sin\phi, \cos\theta)$. The task of a reconstruction algorithm is to provide estimates $\hat{p}_x, \hat{p}_y, \hat{p}_z, \hat{\theta}, \hat{\phi}$ for these five parameters. In addition it must provide information on the "goodness of the fit" and the expected errors on the results.

V. Kulikovskiy, *Neutrino Astrophysics with the ANTARES Telescope*,
Springer Theses, DOI 10.1007/978-3-319-20412-3_4

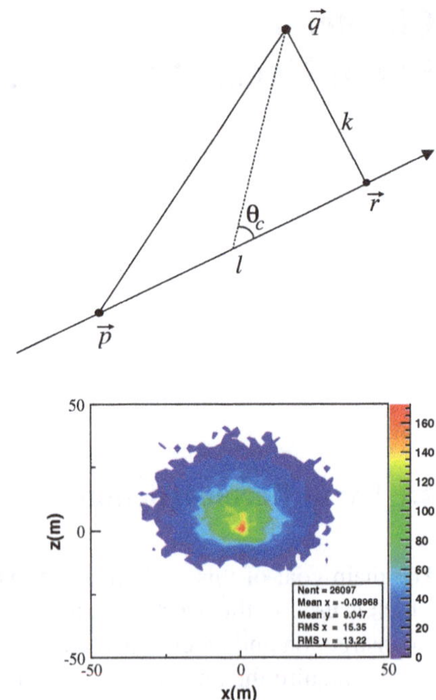

Fig. 4.1 Geometrical relation between a point $\vec{q}$ and the muon track. k and l represent the components of $\vec{v} = \vec{q} - \vec{p}$ perpendicular, and parallel to the track direction respectively. The *dashed line* indicates the path of the light. θ_C is the Čerenkov angle. The image is taken from (Heijboer 2002)

Fig. 4.2 Example of a distribution of the points of closest approach of the muon to a PMT in a coordinate system where the orientation of the PMT is in the positive z-direction. The distribution shown is for hits with an amplitude between 0.5 and 1.5 p.e. The image is taken from (Heijboer 2002)

The arrival time at position $\vec{q}$ of Čerenkov light that is emitted from a muon track can be calculated as:

$$t^{\text{th}} = t_0 + \frac{1}{c}\left(l - \frac{k}{\tan\theta_C}\right) + \frac{1}{v_g}\left(\frac{k}{\sin\theta_C}\right), \tag{4.1}$$

where $l = (\vec{q} - \vec{p}) \cdot \vec{d}$ and $k = \sqrt{(\vec{q} - \vec{p})^2 - l^2}$ as it is shown in Fig. 4.1, θ_C is the Čerenkov angle, v_g is the group velocity of light in the water.

The first stage of the track reconstruction is the "linear prefit" of a straight line through positions in the detector that are associated with the hits. These positions are defined as the points of closest approach of the optical module to the track (point $\vec{r}$ in Fig. 4.1). Simulations show that the distributions of points of the closest approach has the same axial symmetry as OM and the maximum of this distribution is located along the axis of symmetry of the OM (see Fig. 4.2). The (approximate) spherical symmetry of the distribution implies that the uncertainties on the x, y and z-coordinates of the positions are uncorrelated and equally large. They are characterised by the RMS of the distribution. The distance from the closest approach point to the OM and its RMS is shown as function of the hits amplitude in Fig. 4.3. Hits with a high number of photoelectrons correspond to muons passing close to the OM. In conclusion, for each hit its point of the closest approach is estimated as a point along the OM orientation axis located on the distance from the OM centre corresponding to the hits amplitude using the dependence shown in Fig. 4.3 together with the uncertainty of estimation.

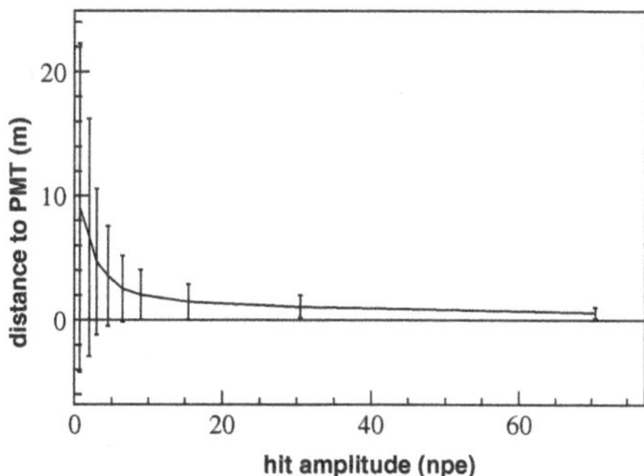

Fig. 4.3 The average distance of the point of closest approach of the muon track to the OM as a function of hit amplitude. The error bars indicate the spread of this position. The image is taken from (Heijboer 2002)

For each hit i with a position $\vec{r}_i$ and time t_i of a hit one can use a simple relation (approximating the point where the muon is located at the moment of the light arrival to the OM with the point of the closest approach to the OM):

$$\vec{r} = \vec{p} + c\vec{d}, \tag{4.2}$$

which for all hits can be summarised as a relation between a vector $\boldsymbol{R} = (x_1, y_1, \ldots, z_n)$ containing all hit positions and a vector $\boldsymbol{\Theta} = (p_x, d_x, p_y, d_y, p_z, d_z)$ containing estimates of the track parameters:

$$\boldsymbol{R} = \boldsymbol{H\Theta}, \tag{4.3}$$

with a matrix $\boldsymbol{\Theta}$ containing hit times:

$$\boldsymbol{H} = \begin{pmatrix} 1 & ct_1 & & & & \\ & & 1 & ct_1 & & \\ & & & & 1 & ct_1 \\ 1 & ct_2 & & & & \\ & & 1 & ct_2 & & \\ \vdots & \vdots & \vdots & \vdots & \vdots & \vdots \\ & & & & 1 & ct_n \end{pmatrix}. \tag{4.4}$$

The errors on the hit positions are stored in the error-covariance matrix, the inverse of which is called $\boldsymbol{F}$. Errors on the hit times are neglected. The estimator for the track parameters $\boldsymbol{\Theta}$ is chosen such that the following quantity is minimised:

$$\chi^2 = (\boldsymbol{R} - \boldsymbol{H\Theta})^T \boldsymbol{F} (\boldsymbol{R} - \boldsymbol{H\Theta}). \tag{4.5}$$

This yields to the following estimation of $\boldsymbol{\Theta}$:

$$\hat{\boldsymbol{\Theta}} = (\boldsymbol{H}^T \boldsymbol{F} \boldsymbol{H})^{-1} \boldsymbol{H}^T \boldsymbol{F} \boldsymbol{R}. \tag{4.6}$$

Assuming that the covariance matrix is diagonal, and if the errors on the x, y and z components of the position of an individual hit are approximately equal, the expression for $\hat{\mathbf{\Theta}}$ simplifies. In the current implementation of the linear fit the velocity of the muon is unconstrained as the length of the vector $\vec{d}$ is unconstrained. The unity condition on $\vec{d}$ may be incorporated using the technique of Lagrange multipliers, although it was seen that the result is identical with the only difference that the $\vec{d}$ is automatically normalised in the last case.

For each hit and a track estimate $\vec{p}, \theta, \phi$ one can calculate a time residual:

$$r_i = t_i - t_i^{\mathrm{th}}(\vec{p}, \theta, \phi), \tag{4.7}$$

where t^{th} is calculated according to (4.1).

The estimates of track parameters are given by those parameters for which the following is maximal:

$$Q = \sum_{i=1}^{N_{\mathrm{hits}}} g(r_i(\hat{\vec{p}}, \hat{\theta}, \hat{\phi})). \tag{4.8}$$

The function $g(r)$ for the next step after linear prefit is chosen such that it describes the data for small residuals. The behaviour for large r should be such that there is a favourable trade-off between an accurate description of the data and the ease of finding the global maximum. The latter depends on the quality of the a priori estimate of the track parameters. After several optimisations, the function to be minimised is chosen to be:

$$Q = -\sum_{i=1}^{N_{\mathrm{hits}}} \alpha \left(2\sqrt{1 + A_i r_i^2/2}\right) + (1-\alpha) f_{\mathrm{ang}}(a_i), \tag{4.9}$$

where A_i is the amplitude of the hit i in p.e., r_i is in ns, and a_i is the cosine of the angle between the photon and the direction of sight of the PMT. The first part of the function describes the behaviour of the time residual close to the minimum and this function was chosen to be flatter than the classical $-r^2$ to be more robust to not well defined starting tracks from the linear fit. The second part contains the angular response function $f_{\mathrm{ang}}(a_i)$ of the optical module shown in Fig. 3.10, which describes the fact that the OM should be able to see the light from the track. The proportion between the two parts is defined by α for which the chosen value is $\alpha = 0.05$. The small value of α does not mean that the second term is dominant in the fit, since the influence of the two terms depends on their derivatives with respect to the track parameters.

At the next step a more sophisticated $g(r)$ function which better describes the data may be used. Logarithm of the simplified likelihood function $\mathcal{L}(r)$ obtained from the simulations was used for this purpose. This method estimates the parameters of the track by finding the track parameters that yield the highest probability of the observed data. The likelihood function is compared in Fig. 4.4 with $-r^2$ function and the first part of the function (4.9).

Finally, for the last track estimator a complete likelihood function is used. It depends on the time residuals, the estimated travelled distance of the photons, estimated angle between the arrived photons and the OMs and the optical noise background. Its parametrisation may be found in the former work (Heijboer 2004, Sect. 4.4).

The combination of the fitting algorithms into one reconstruction algorithm is schematically shown in Fig. 4.5. The first step is a rough selection of the signal hits: all hits that fulfil the causality criterion (2.2) with the largest hit that is part of a coincidence are selected. The Linear prefit is applied on a high purity subset of these hits. This subset consists of all hits that are part of a coincidence and all hits that have an amplitude of at least 3 p.e. Few of these hits are background hits.

From the result of the prefit, nine different tracks are generated slightly varying prefit track direction and position. These are used as starting points for the next procedure. For each of the starting points, compatible hits are selected from the causality connected hits with the time residuals less than 150 ns and a distance to the track less than 100 m. In the next stage the M-estimator fit (4.9) is applied to these hits, followed by a simple likelihood fit. This fit improves the accuracy of the result of the M-estimator fit. The solution with the highest likelihood per number degrees of freedom ($N_{\text{hits}} - 5$) is considered the best one. In addition "secondary" solutions

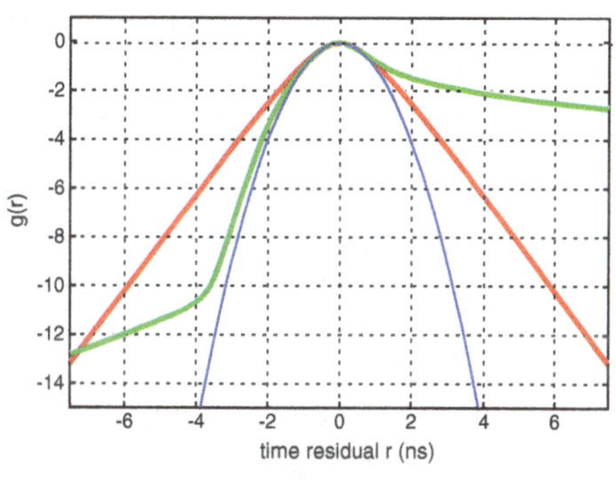

Fig. 4.4 Comparison of the fitting functions: $-r^2$ (*blue*), $-2\sqrt{1+r^2/2}+2$ (*blue*), simplified likelihood function $\ln \mathcal{L}(r)$ (*green*). The image is taken from (Heijboer 2002)

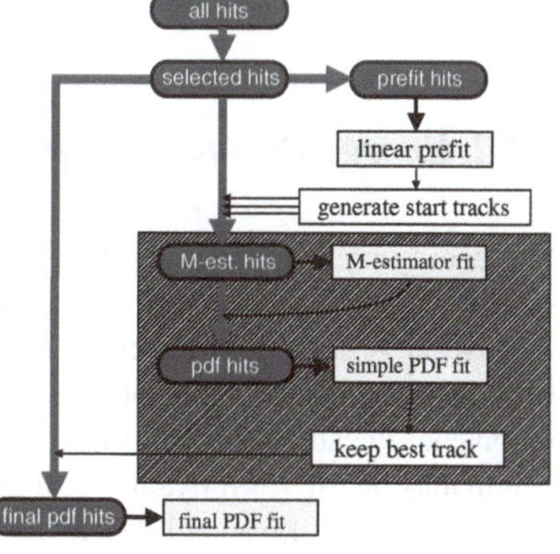

Fig. 4.5 Schematic depiction of the reconstruction algorithm. The image is taken from (Heijboer 2002)

may be present; they are stored for later use. The best track is used as a starting point for the final fitting process which consists of a fit with the improved likelihood. The cut on the residuals and the distance of the hits used for the final fit is loose: 250 ns and 300 m respectively. Hence it allows for background hits to be present. Finally, error estimates are computed for the fitted track varying the parameters till the likelihood increases by 1/2 (some explanations of "$\mathcal{L} + 1/2$" method are done in Sect. 5.5.1, more details are in (Casella and Berger 2001).

The final step is a likelihood maximisation. The likelihood is defined using the probability density function for the observed hits and a track. The initial steps—linear χ^2 fit and more simplified time residuals fits, are needed to find an input track for the final fit and to select photomultipliers close to this track in space and time. It was found that the likelihood function for the final fit has many local maxima. This requires initial track needs to be close to the true one. Moreover, to increase the chance to find a correct maximum an initial track for the final fit is slightly changed to get nine different starting tracks. The final track is a track with the maximum likelihood and the quality of the reconstruction is estimated by:

$$\Lambda = \frac{\mathcal{L}}{N_{\text{hits}} - 5} + 0.1 \times (N_{\text{comp}} - 1), \tag{4.10}$$

where N_{hits} is the number of hits used in the fit. This quality parameter incorporates the maximum value for the likelihood, $\mathcal{L}$, per number of degrees of freedom and the number of initial tracks converged to the same final track N_{comp}. In general, N_{comp} is one for the badly reconstructed events and goes till nine for the well reconstructed events.

Also an uncertainty on the track direction β is estimated. Neutrino simulations with E^{-2} spectrum was used to estimate the angle between the direction of the reconstructed muon and that of the true neutrino (Adrián-Martínez et al. 2012). Of the events selected with $\Lambda > -5.2$ and $\beta < 1°$, 83 % are reconstructed better than 1°.

Shower-like events are identified by using a second tracking algorithm with χ^2-like fit, assuming the hypothesis of a relativistic muon (χ^2_{track}) and that of a shower-like event (χ^2_{point}) (Aguilar et al. 2011). Events with better point-like fit ($\chi^2_{\text{point}} < \chi^2_{\text{track}}$) have been excluded in this work.

4.1.2 Energy Reconstruction

The energy of muons can not be estimated precisely from the detected light in the ANTARES detector due to the limited size of the instrumented volume. The energy estimation is done on an average base of a large number of events. For the tracks fully contained in the detector sensitive volume the track length may be used for the muon energy estimation. For the high energies the amount of detected light per track length may be used as the radiative losses are proportional to the energies of the muon according to Fig. 4.6 and (1.21). Moreover, radiative processes produce hadronic and

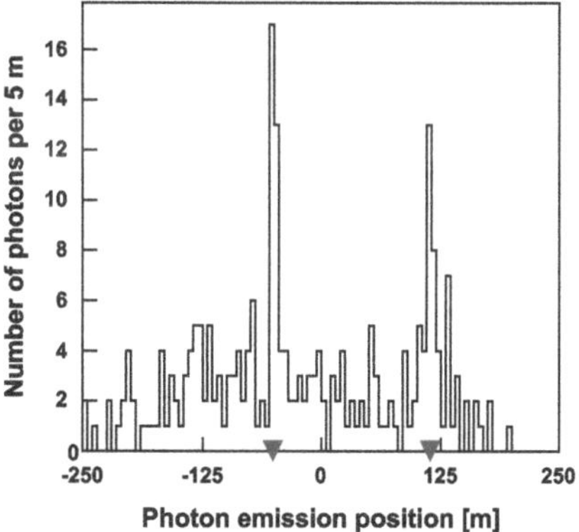

Fig. 4.6 Example of the number of detected photons projected along the muon trajectory for the event in the ANTARES data. Two showers are seen (*triangles*). This image is taken from (Aguilar et al. 2012)

electromagnetic cascades for which the light emission direction is different from the Čerenkov cone of the primary track. These cascades may be seen in the ANTARES data by projecting positions of the detected photons along the reconstructed track (Fig. 4.6). Together with scattering in water this effect produces delayed photons in respect to the Čerenkov photons. The number of delayed photons may serve as another parameter for the energy estimation.

In this analysis the energy was estimated using Artificial Neural Networks (ANNs), which are produced using a machine learning algorithm which derives the dependence between a set of observables and the energy estimate in a semi-parametric way (Schnabel 2012). A simple ANN is made up of nodes, similar to the brain cells of the human brain, in which a decision is made, and connections between those nodes. The connections pass signal between the nodes with a strength that is adapted through training and resemble the synapses in the brain.

In each node j, the input parameters x_i entering from the connected nodes are signed up with assigned connection weights w_{ij}. The resulting input then triggers the reaction of the node, which is calculated using the activation function g. The activation function can vary between a step-function, which results in a yes/no decision, and continuous function, which allows for modelling of a sophisticated functional dependence. Thus an ANN can be used to map both linear and non-linear problems. The output z_j of each node is calculated by:

$$z_j = g\left(\sum_i w_{ij} x_i\right). \tag{4.11}$$

The activation function used in this woks is $g(x) = \tanh(x/2)$ which is antisymmetric and its slope close to unity at 0. ANN normally deals with values within $[-1, 1]$, so all input parameters were normalised and span in this range using trigonometric functions. For the output inverse procedure was adopted to obtain energy in GeV.

In the feed-forward networks which are used in this work the nodes are arranged in layers. Nodes handling input parameters from outside the ANN are called input nodes and they are forming input layer. Each node receives input from all nodes of

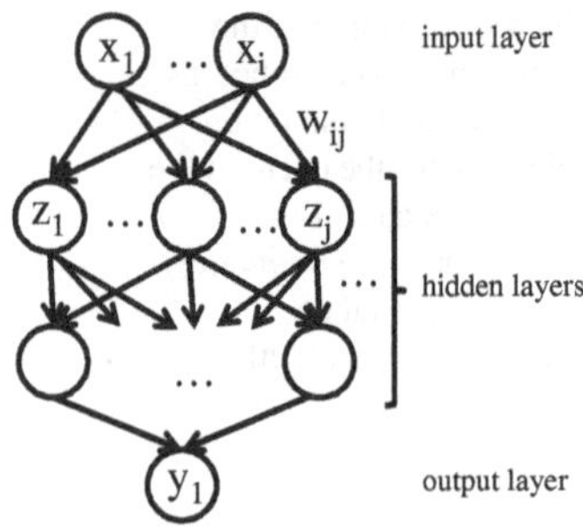

Fig. 4.7 Schematic view of an ANN with *circles* indicating the nodes, the connection weights w_{ij}, input parameters $x_1 \dots x_j$ and output y_1

the previous layer and passes its output to all nodes in the following layer. Layers which have no direct connection to the outside but only to other nodes are called hidden layers. The scheme is shown in Fig. 4.7. Number of hidden layers used in this works is two. There is one output node y_1 used in this scheme as the aim of the ANNs is to estimate just one variable (energy).

The adaptivity of the ANN to a given problem is provided by the variability of the connection weights. During training, a fixed set of input and output values is presented to the ANN using simulated events. The input is propagated through the ANN, at first with randomised connection weights, and compared to the output of the training set. For the complete set with N_{sample} samples, the Mean Square Error (MSE) is calculated as:

$$\text{MSE} = \frac{1}{N_{\text{sample}}} \sum_{k=1}^{N_{\text{sample}}} (y_k^{\text{ANN}} - y_k^{\text{true}})^2, \tag{4.12}$$

The connection weights are now adapted recursively, starting from the output layer, to minimise the MSE between the training sample and the ANN output using a backpropagation algorithm (Haykin 1999). One complete run through the data set and adapting of the weights is called training epoch. In order to avoid overfitting, the MSE of an independent set of data (validation set) is evaluated after each epoch. If the MSE of the validation set increases, the ANN starts to adapt to the special features of the training set and the training is stopped (early stopping method). Each simulated data set is divided into training set and validation set with ratio 2:1 in this work. An example of MSE evolution while training is shown in Fig. 4.8. When the MSE of the validation set rises in 5 consecutive epochs, the training is stopped.

When selecting the input parameters to the ANN, two criteria are of importance: on the one hand the input parameters should be simple, as all abstraction is done within the ANN, on the other hand the parameters should span the relevant feature space as widely as possible. In ANTARES, this means in the simplest case deducing relevant input parameters from the timing and location of the photons in the detector. If one goes one step further, a muon track has already been reconstructed and the parameters of the reconstructed track can be assumed as given.

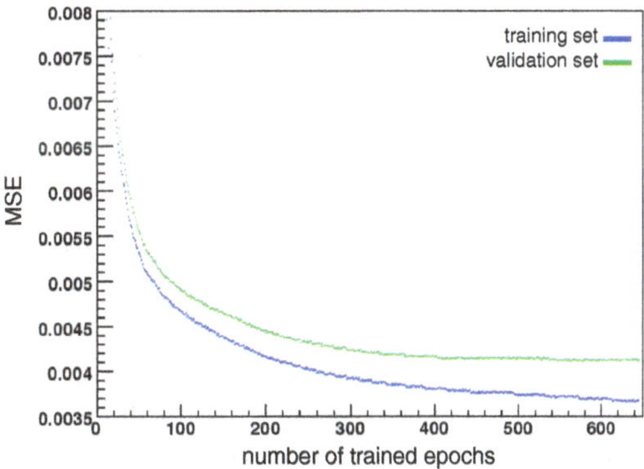

Fig. 4.8 MSE of the training and validation set for each training epoch

For the hit selection when no track has been yet reconstructed, the first task is to find hits which might be causally connected to the muon. The hits which were found by the trigger algorithm (L2 hits) provide a good starting point for this, but offer only a very strict selection. To include almost all hits which origin from the muon, all hits within a time window between 20 ns before the first triggered hit time and 300 ns after the last triggered hit time were selected, knowing that this selection will inevitably also include background hits. To offer the ANN an estimate of the background from bioluminescence, the number of hits in two time windows 1 μs before and after the triggered hits were also added to the input parameters. Parameters without muon track which were used in this work are:

Hits and charge information The most straight-forward observable connected to the energy is the number of hits and the hit charge measured by the detector. Also the mean and RMS of the distribution of these observables were added to the list, as they might indicate the presence of showers along the muon track.

Lines and storeys information Number of detector lines and storeys, which have L2 hits is calculated.

Timing information Without track reconstruction, parameters which include the time information of the hits are only very rough estimates. The time between the first and last triggered hit is an obvious choice of parameter. In order to describe the temporal distribution of the hits, the timing of hits in reference to the first triggered hit was used and added to the parameter list through its mean and RMS.

Parameters which use a muon track reconstructed as it described before are:

Analysing hit distributions Hit amplitude, timing and distance distributions in reference to Čerenkov photons originating from the track were investigated and their mean and RMS added to the parameter list. The effective track length of the muon was also included, meaning the distance along the muon track between the projection points of the first and last triggered hit. Considering the readout mode of the OMs, hits were also divided into those measured within 40 ns from the direct Čerenkov photons and those arriving later, and their number and amplitude added.

Non-linear approach: Cylinder-counts In reference to the track, the detection volume was divided into cylindrical shells with a thickness of 20 m. Within each shell, the number of working OMs, the number of hits and total amplitude in different time windows and the acceptance of the OM for Čerenkov photons is calculated. This expands the list to over 50 parameters.

The large number of input parameters might be surprising at first, but can be explained in two ways: on the one hand, the parameters are decoupled through a Principle Component Analysis (PCA) before being passed to the ANN, which emphasises the common features in the parameters. PCA is mathematically defined (Jolliffe 2002) as an orthogonal linear transformation that transforms the data to a new coordinate system such that the greatest variance by any projection of the data comes to lie on the first coordinate (called the first principal component), the second greatest variance on the second coordinate, and so on. On the other hand the parameters draw on two different functionalities of the ANN. The continuous parameters aim at the function approximation capability of ANNs, while the cylinder parameters use their decision-making functionality. The parameters are therefore not competitive, but complementary. The large number of parameters can also make the ANN more stable against small divergences of the data from the training Monte Carlo in one parameter, as it does not have to rely on one parameter alone, but on features spanning a much larger range.

To increase the accuracy of the ANN, the real data from the detector without triggered events is used to simulate the background. This data is divided into runs with a mean rate below 75 kHz, between 75–100 kHz, 100–150 kHz and above 150 kHz. Also the runs are divided according to the number of active OMs, resulting in specialised ANNs for 5, 9, 10 and 12 active detector strings. Therefore, overall 15 different ANNs are trained (there was not enough statistics for 12 lines with mean rate above 150 kHz).

The training set should include muons at almost all energies to some extent, as well as all zenith angles, to achieve a representation of all relevant features. An approximately even distribution in logarithmic energy and zenith angle ensures that no bias is introduced towards an over-represented energy. For the data set, a total of about 300,000 events was used to cover the complete available energy range between 10 GeV to 100 PeV. To account for the larger probability of downgoing atmospheric muons compared to upgoing neutrino candidates, roughly 1/3 of the sample is made of atmospheric muons, 1/6 of downgoing neutrinos and the remaining 1/2 of upgoing neutrinos. After dividing the data set into training and validation set and cutting the sets such that the logarithmic energy was distributed evenly within the sets, about 50,000 events remained in the training sample.

The median resolution for $\log_{10} E_{\mathrm{Rec}}$ is shown in Fig. 4.9 and it is about 0.3 for muons with an energy of 10 TeV. The agreement between data and simulation for the reconstructed energy of the atmospheric neutrinos will be shown in Sect. 5.4.

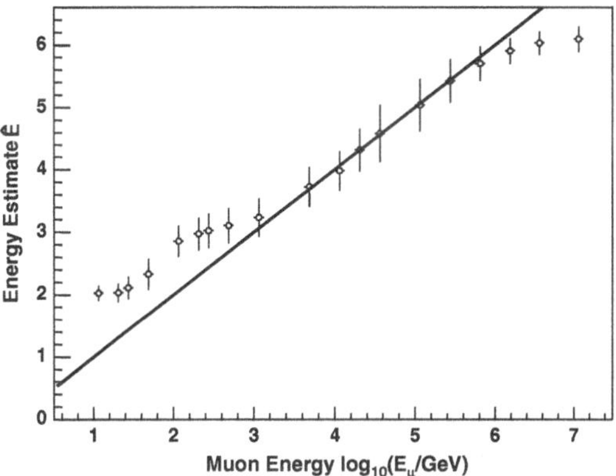

Fig. 4.9 Energy estimates from ANN for repeated simulations of upgoing muon tracks at discrete energies. The image is taken from (Schnabel 2013)

4.2 Celestial Coordinates and Maps

Track reconstruction in local coordinate system of the detector together with the time information of the event allows to evaluate the position of neutrino origin on the sky. Astronomers use different celestial coordinates systems and maps. Coordinates and maps used in this work are described in the following sections.

4.2.1 Celestial Coordinates

There are two important celestial coordinate systems, used in this thesis: galactic and equatorial.

Galactic coordinates uses the Galactic Centre as its centre, the fundamental plane approximately in the galactic plane and the primary direction aligned with the approximate centre of the Milky Way galaxy. It uses the right-handed convention, meaning that coordinates are positive toward the north and toward the east in the fundamental plane. The precise definition of the galactic coordinate reference points is defined through the equatorial coordinates.

The equatorial coordinate system is centred at Earth's centre, but fixed relative to distant stars and galaxies. The coordinates are based on the location of stars relative to Earth's equator if it were projected out to an infinite distance. A fundamental plane consists of the projection of the Earth's equator onto the celestial sphere (forming the celestial equator), a primary direction towards the vernal equinox, and a right-handed convention. In order to fix the exact primary direction, these motions necessitate the specification of the equinox of a particular date, known as an epoch, when giving a position. In the current work a mean equinox of a J2000.0 standard epoch is used. It is a fixed standard direction, allowing positions established at various dates to be compared directly. Two coordinates in this system: declination and right ascension defines as following. Declination measures the angular distance of an object perpendicular to the celestial equator, positive to the north, negative to the south. Right ascension measures the angular distance of an object eastward along the celestial equator from the vernal equinox to the hour circle passing through the object.

4.2.2 Sky Maps

The problem for sky maps originates from the common problem of representing a sphere on a plane (it can be the Earth sphere or celestial sphere for galactic or equatorial coordinates). And there is no standard solution. It is always a compromise of what parameters we want to represent better.

There are several properties which projections may have (not all of them are listed here):

- A projection is said to be equidistant if the meridians are uniformly, truely, or correctly divided so that the parallels are equispaced. That is, the native latitude is proportional to the distance along the meridian measured from the equator, though the constant of proportionality may differ for different meridians. Equidistance is not a fundamental property. Its main benefit is in facilitating measurement from the graticule since linear interpolation may be used over the whole length of the meridian. This is especially so if the meridians are projected as straight lines which is the case for all equidistant projections presented here.
- Certain equal-area projections (also known as authalic, equiareal, equivalent, homalographic, homolographic, or homeotheric) have the property that equal areas on the sphere are projected as equal areas in the plane of projection. This is obviously a useful property when surface density must be preserved.
- Zenithal, or azimuthal projections give the true azimuth to all points on the map from the reference point at the native pole (Calabretta and Greisen 2002).

In astrophysics one of the most popular projection is the so-called "Aitoff projection" (also noted as "Aitoff-Hammer" or "Hammer-Aitoff"). In reality, most astrophysicists miscalled Hammer projection as Aitoff projection. Actually, they look very similar. The comparison between the grid lines for both projections may be seen in Fig. 4.10. (Hammer is black and Aitoff is red). Aitoff projection is the equa-

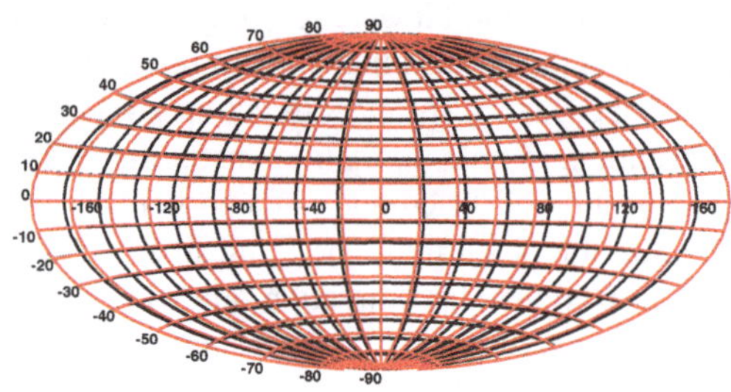

Fig. 4.10 Aitoff (*red*) and Hammer projection (*black*) grid lines

torial form of the azimuthal equidistant projection. So, it saves angles, the property, which is very important for presenting the positions of sources. Hammer projection, instead, is equatorial form of the Lambert azimuthal equal-area projection. So, it saves area which is very important for the representing fluxes, anisotropies and etc.

The Hammer projection was described by Ernst Hammer in 1892. Using the same 2:1 elliptical outer shape as the Mollweide projection, Hammer intended to reduce distortion toward the outer limbs, where it is extreme in the Mollweide. Directly inspired by the Aitoff projection, Hammer suggested the use of the equatorial form of the Lambert azimuthal equal-area projection instead of Aitoff's use of the azimuthal equidistant projection. The exact equations may be found here (Calabretta and Greisen 2002, p. 1094).

Also for the sky map representation, the minimum and maximum of the axis as well as their direction differs. It is rather common to see the altitude or declination varying from $-90°$ at the bottom to $90°$ at the top of the map. While, for the longitude or right ascension different combinations maybe found in literature: from $0°$ to $360°$ or from $-180°$ to $180°$ with the increase from left to right or opposite. For the galactic coordinates I have chosen the axis from $180°$ to $-180°$ (left to right) and for equatorial coordinates from $0°$ to $360°$ which correlates with the previous ANTARES publications.

4.3 Data Quality

The ANTARES detector is collecting the data in a semiautomatic way. Every run change the configuration of the completed run is reused if there is no interruption from the detector operators crew. The data taking runs smooth except some rare cases like a power outage or a lost of the connection to the DataBase. Also the connection with some optical modules can be lost at the run start or during the data taking. The data quality control serves to exclude the corrupted data from the physics analysis. The data quality is divided into two levels. On the low level simple parameters of the run are checked, like run duration comparison with number of the data time slice. On the high level the quality of the runs involving the event reconstruction and comparison with simulated data is performed.

4.3.1 Low Level Data Quality

Several quantities are checked for each run:

- minimal duration: 1 second (as computed from the effective number of data time slices in the run).
- at least one active ARS (i.e. at least one data frame of 104.8 ns with the rate >10 kHz)

- no synchronisation problems:
  - no more than 2 % of data time slices with an offset between their start time and the time slice start signal arriving form the master clock
  - no more than 2 % of data time slices with data frames exhibiting a mismatch between their time and their time slice index (each data frame has an index of the time slice to which they belong, the time slice index is increased by one for the next time slice)
  - index of last time slice roughly matches total number of slices (taking into account the time slice sampling): Slicecount − (Nslices × Sampling) < 100; sampling is used to reduce the data flow during the high optical background periods
  - time of the data frame start in the DataBase matches the frame time stored in the data
- no double frames: no ARS producing more than one frame in a given time slice
- sampling of time slices <3, the higher sampling corresponds to extremely noisy periods
- limited time lost during the run: $0 \leq \text{TotalTime} - \text{EffectiveTime} \leq 450$ s, where $\text{TotalTime} = T_{\text{stop}} - T_{\text{start}}$ and $\text{EffectiveTime} = \text{Nslices} \times \text{FrameTimeDuration} \times \text{Sampling}$
- trigger rate within acceptable limits: $10\ \text{mHz} \leq R \leq 100\ \text{Hz}$

The total data rejected for further analysis due to each criteria is shown in Fig. 4.11. As it can be seen, most problems disappeared after the short period from the detector deployment.

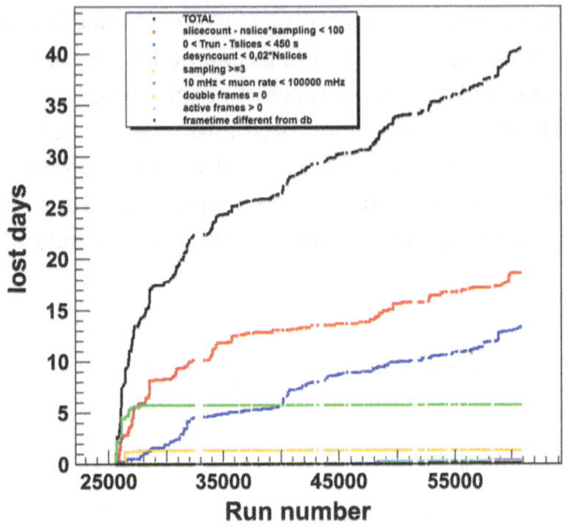

Fig. 4.11 Rejected ANTARES data runs for the data taking period 2007–2011

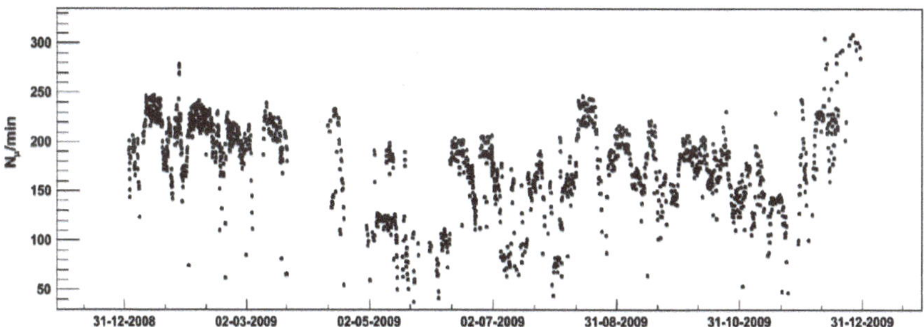

Fig. 4.12 Average muon rate per run in the ANTARES detector in 2009

4.3.2 Data Quality Using Monte Carlo Simulation

The simple idea for this task is to check the number of the atmospheric muons in the data. The atmospheric muon flux can be assumed constant on level of 1 % due to the temperature variations in the atmosphere. This allows the data check at the event reconstruction level which includes time, charge and position calibration.

The muon rate in the ANTARES detector is shown in Fig. 4.12. Unlike the naively expected constant muon rate, one can see the variations up to factor of four, which are caused by the different detector configuration including triggers used, environmental conditions and number of active photomultipliers. The last one is intuitively the most efficient in the muon reconstruction. In particular, due to the bioluminescence activity in the detector environment, even if during a single run the number of working PMTs is constant, not all the data from them are available offshore due to HRV and Xoff explained in Sect. 3.3.4.

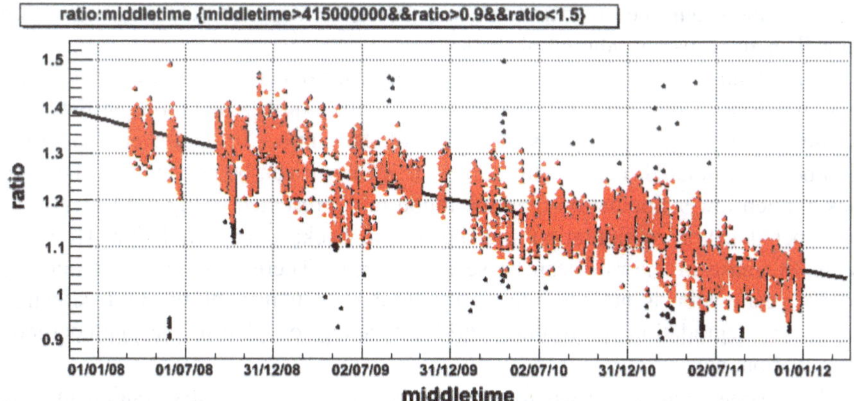

Fig. 4.13 Ratio of the data events to the simulated ones in one run with $\Lambda > -6.5$ (*black dots*) versus time of the run start with the fit (*black line*). *Red dots* correspond to the selection $|\log_2(\text{ratio}/\text{fit})| < 3\,\sigma$

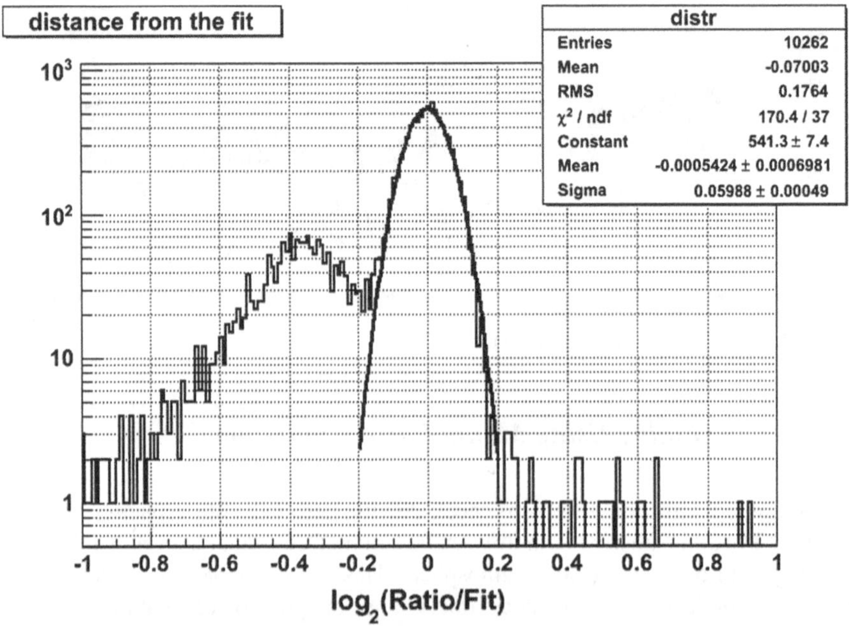

Fig. 4.14 Distribution of the ratio of the data events to the simulated ones in one run with $\Lambda > -6.5$

Initially, the method, based on comparison of the average muon rate and the average number of active PMTs was used for the data control (Gracheva et al. 2011). In particular, it was shown that the logarithm of muon rate linearly depends from the number of the active PMTs for the similar detector configuration. However, this method has a difficulty for the cases when the detector configuration is changing significantly (new lines connected/disconnected, trigger setup is changed and etc.). So, the runs should be organised in groups with the similar detector configuration first. This requires fair amount of work.

For the run by run simulation approach when each data run has its own simulation it is possible to make a direct comparison of the number of muons in data and in simulation. Figure 4.13 shows the ratio of the number of data events to to the number of simulated muons with $\Lambda > -6.5$. First it can be noticed that the ratio has a much lower fluctuation than the muon rate, presented in Fig. 1.14, thanks to a proper detector configuration and the optical background simulation. However some decrease of the ratio can be seen over time. There is about 6 % efficiency loss a year which is not reproduced in simulation. It may be the OM detection efficiency degradation (biofouling, PMT efficiency loss due to the photocathode degradation, etc.).

The exponential fit is used for the number of events ratio distribution in time (as a fixed percent of worsening in time is expected). For the following analysis it was decided to remove the runs whose ratio stays away from the fit. Basically,

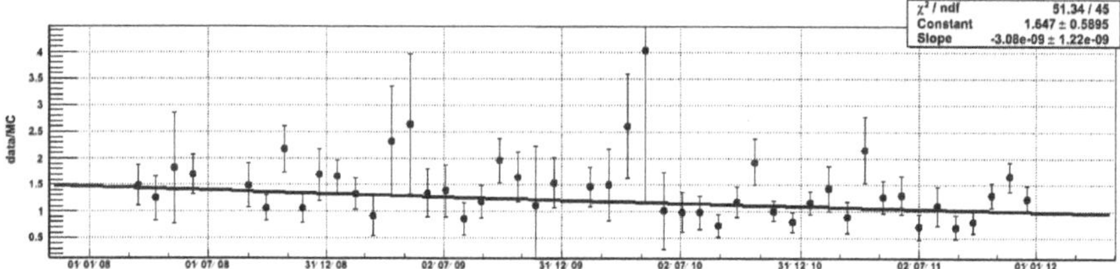

Fig. 4.15 Ratio of the upgoing data events to the simulated ones in one run with $\Lambda > -5.1$ coming from the off-zones

the $\log_2(\text{ratio}/\text{fit})$ for each run was calculated. The distribution was fitted with a Gaussian function as it is present in Fig. 4.14. It was decided to keep runs with $|\log_2(\text{ratio}/\text{fit})| < 3\,\sigma$.

At the end 8193 runs were selected which corresponds to 806 days from 888 days of the data available. Also for these runs grouped by 4 calendar weeks a ratio of the number of upgoing events from the offzones in the data to the simulated ones with $\Lambda > -5.1$ was calculated (the choice of the cut and the off-zones selection will be explained later). Similar decrease may be seen in the ratio presented in Fig. 4.15 although the fit values are consistent with the no decrease hypothesis on a $2\,\sigma$ level.

References

S. Adrián-Martínez et al., Search for cosmic neutrino point sources with four years of data from the ANTARES telescope. ApJ **760**(1), 53 (2012). ISSN: 0004–637X, http://stacks.iop.org/0004-637X/760/i=1/a=53?key=crossref.6fb56c5cd11c693e9cebcb0f5c28eb25

J. Aguilar et al., A fast algorithm for muon track reconstruction and its application to the ANTARES neutrino telescope. Astropart. Phys. **34**(9), 652–662 (2011). ISSN: 09276505, http://linkinghub.elsevier.com/retrieve/pii/S0927650511000053

J. Aguilar et al., A method for detection of muon induced electromagnetic showers with the ANTARES detector. NIM A **675**, 56–62 (2012). ISSN: 01689002, http://linkinghub.elsevier.com/retrieve/pii/S0168900212001179

M.R. Calabretta, E.W. Greisen, Astrophysics representations of celestial coordinates in FITS. Astron. Astrophys. **395**, 1077–1122 (2002)

G. Casella, R. Berger, *Statistical Inference*, 2nd edn. (Duxbury, Pacific Grove, 2001). ISBN: 0-534-24312-6, http://departments.columbian.gwu.edu/statistics/sites/default/files/u20/Syllabus6202-Spring2013-Li.pdf

K. Gracheva et al., Down going muon rate monitoring in the ANTARES detector. SIDSPubblicazioni (INFN/TC-11/05) (2011). http://www.lnf.infn.it/sis/preprint/pdf/getfile.php?filename=INFN-TC-11-5.pdf

S. Haykin, *Neural Networks: A Comprehensive Foundation*, vol. 13 (Prentice Hall, Upper Saddle River, 1999)

A. Heijboer, An algorithm for track reconstruction in ANTARES. Technical report. ANTARES-SOFT-2002-002 (2002)

A. Heijboer, Track reconstruction and point source searches with ANTARES, Ph.D. thesis, Institute for High Energy Physics (IHEF), 2004. http://dare.uva.nl/document/77462

I.T. Jolliffe, *Principal Component Analysis*, 2nd edn. (Springer, 2002). ISBN: 978-0-387-22440-4

J. Schnabel, The ANNergy estimator; using artificial neural networks for muon energy reconstruction. Technical report. ANTARES-SOFT-2012-010 (2012)

J. Schnabel, Muon energy reconstruction in the ANTARES detector. NIM A **725**, 106–109 (2013). ISSN: 01689002, http://linkinghub.elsevier.com/retrieve/pii/S0168900212016658

Chapter 5
A Search for Neutrino Emission from the Fermi Bubbles

Analysis of the Fermi-LAT data has revealed two extended structures above and below the Galactic Centre, emitting gamma rays with a hard spectrum, the so-called Fermi bubbles. Hadronic models attempting to explain the origin of the Fermi bubbles predict the emission of high-energy neutrinos and gamma rays with similar fluxes. In this thesis a search for neutrinos from the direction of the Fermi bubbles using ANTARES data has been performed. The search was done using a blinded analysis which excludes possible biases during the optimisation of the signal to noise ratio and, therefore increases the reliability of the measurement. Thus, the data in the area of the Fermi bubbles have been blinded before the precise analysis scheme has been established and approved by the collaboration.

The hypothetical neutrino emission from the Fermi bubbles is described in Sect. 5.1. Then, the search for neutrino emission is performed by comparing the number of events in the Fermi bubble regions to the number found in similar off-zone regions (Sect. 5.2). The event selection optimisation is based on a simulation of the expected signal as described in Sect. 5.4. Finally, the selected events are presented in Sect. 5.5 together with the significance and the upper limit on the neutrino flux from the Fermi bubbles.

5.1 Fermi Bubbles

Analysis of data collected by the Fermi-LAT experiment has revealed two large circular structures in our Galactic Centre and perpendicular to the galactic plane—the so-called Fermi bubbles (Su et al. 2010). The total size of the gamma-ray emission area is about 0.8 sr. The approximate edges of the Fermi bubble regions are shown in Fig. 5.1.

These structures are characterised by gamma-ray emission with a hard E^{-2} spectrum and a constant intensity over the whole emission region. One of the energy spectrum fits is shown in Fig. 5.2. Evaluation of the spectrum depends on the template for

V. Kulikovskiy, *Neutrino Astrophysics with the ANTARES Telescope*,
Springer Theses, DOI 10.1007/978-3-319-20412-3_5

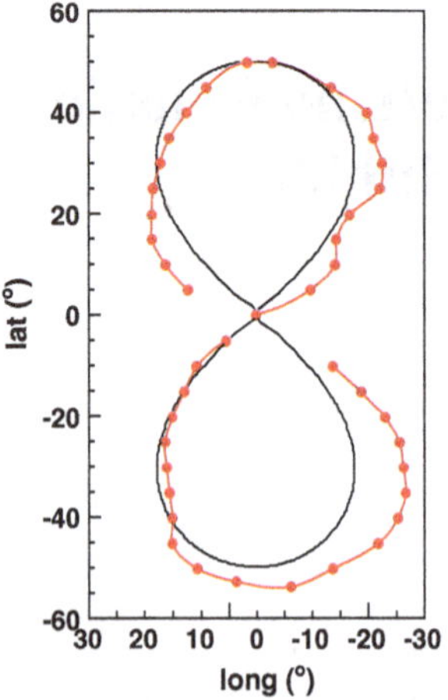

Fig. 5.1 Approximate edges (*red line, circles*) of the north and south Fermi bubbles respectively in galactic coordinates identified from the 1–5 GeV maps built from the Fermi-LAT data (Su et al. 2010). The *contour line* is discontinuous at the region of the Galactic Centre as the maps are severely compromised by the poor subtraction and interpolation over a large number of point sources in this region. The simplified shape of the Fermi bubbles used in this analysis (*black line*) is also shown

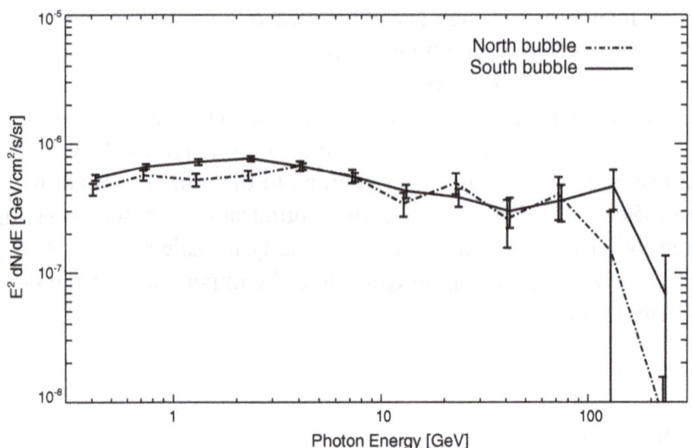

Fig. 5.2 Estimated energy spectra using the Haslam 408 MHz map (Haslam et al. 1982) for the IC (and to a lesser degree bremsstrahlung) emission from SN-shock-accelerated electrons. The image is taken from Su et al. (2010)

the Fermi bubbles shape and the background estimation. The estimation uncertainty is included later in (5.1).

Signals from roughly the Fermi bubble regions were also observed in the microwave band by WMAP (Dobler 2012) and, recently, in the radio-wave band (Carretti et al. 2013). Moreover, the edges correlate with the X-ray emission measured by ROSAT (Snowden and Egger 1997). Several proposed models explaining the emission include hadronic mechanisms, in which gamma rays together with neutrinos are produced by the collisions of cosmic-ray protons with interstellar matter (Crocker and Aharonian 2011; Lacki 2013; Thoudam 2013). Other which include leptonic mechanisms or dark matter decay exclude the neutrino emission or yield lower fluxes (Chernyshov et al. 2011; Dobler et al. 2011; Lacki 2013; Mertsch and Sarkar 2011; Su et al. 2010). The observation of a neutrino signal from the Fermi bubble regions would play an unique role in discriminating between models.

The estimated photon flux in the energy range 1–100 GeV covered by the Fermi-LAT detector from the Fermi bubble regions is (Su et al. 2010):

$$E^2\frac{\mathrm{d}\Phi_\gamma}{\mathrm{d}E} \approx 3-6\times 10^{-7}\,\mathrm{GeV\,cm^{-2}\,s^{-1}sr^{-1}}, \tag{5.1}$$

where the uncertainty is due to background model and Fermi bubbles shape choices. Assuming a hadronic model in which the gamma-ray and neutrino fluxes arise from the decay of neutral and charged pions respectively, the ν_μ and $\overline{\nu}_\mu$ fluxes are proportional to the gamma-ray flux with proportionality coefficients of 0.211 and 0.195 respectively (Villante and Vissani 2008). With this assumption and using (5.1) the expected neutrino flux is:

$$E^2\frac{\mathrm{d}\Phi_{\nu_\mu+\overline{\nu}_\mu}}{\mathrm{d}E} \approx 1.2-2.4\times 10^{-7}\,\mathrm{GeV\,cm^{-2}\,s^{-1}sr^{-1}} \equiv A_{\mathrm{theory}}. \tag{5.2}$$

The neutrino flux, as well as the gamma-ray flux, is expected to have an exponential energy cutoff, so the extrapolation of (5.2) towards higher energies can be represented by:

$$E^2\frac{\mathrm{d}\Phi_{\nu_\mu+\overline{\nu}_\mu}}{\mathrm{d}E} \approx A_{\mathrm{theory}}e^{-E/E_\nu^{\mathrm{cutoff}}}. \tag{5.3}$$

The cutoff is determined by the primary protons which have a suggested cutoff E_p^{cutoff} in the range from 1 to 10 PeV (Crocker and Aharonian 2011). The corresponding neutrino-energy cutoff may be estimated by assuming that the energy transferred from p to ν derives from the fraction of energy going into charged pions (∼20 %) which is then distributed over four leptons in the pion decay. Thus:

$$E_\nu^{\mathrm{cutoff}} \approx E_p^{\mathrm{cutoff}}/20, \tag{5.4}$$

which gives a range from 50 to 500 TeV for E_ν^{cutoff}.

5.2 Off-Zones for Background Estimation

The simplified shape of each Fermi bubble as used in this analysis is shown in Fig. 5.1. The lemniscate of Bernoulli was used to describe their common border. The area can be expressed with the following equation in the Galactic coordinates:

$$(\text{longitude}^2 + \text{latitude}^2)^2 \leq 50^\circ(\text{latitude}^2 - \text{longitude}^2). \quad (5.5)$$

The simplified shape has an angular area of 0.66 sr in comparison to 0.83 sr for the size of the original area. The maximum angular distance from the original border points to the simplified border is 7.7°, the median is 2.7°.

A signal from the combined Fermi bubble regions is searched for by comparing the number of selected events from the area of both bubbles (on-zone) to that of similar regions with no expected signal (off-zones).

Off-zones with the same size and shape as the on-zone are defined using the same solid angle in local coordinates as the on-zone events, but shifted with some fixed delay in time. This ensures the same expected number of background events since it depends on the efficiency of the detector, which is a function of the local coordinates only. The off-zones defined in this way are fixed in the sky.

The size of the Fermi bubbles allows at maximum three non overlapping off-zones to be selected. The on-zone and three off-zones are shown in Fig. 5.3 together with the sky visibility. The visibility of each point on the sky is the fraction of the sidereal day during which it is below the horizon at the ANTARES site (in order to produce upgoing events in the detector). The average visibility of the Fermi bubbles is 0.68 (0.57 for the northern bubble and 0.80 for the southern bubble) and it is the same for the off-zones.

Slightly changing detector efficiency with time and gaps in the data acquisition can produce differences in the number of background events between the on-zone and the three off-zones. For the final event selection it is important to have the same expected number of background events in the on-zone and the off-zones. Event selection described in Sect. 5.4 is done with two parameters: the track quality Λ (4.10) and the reconstructed muon energy E_{Rec} (Sect. 4.1.2). Such an effect was tested in the following way: firstly, the number of events in the off-zones is extracted from the data for various cuts on (Λ, E_{Rec}). The difference in the event numbers between each pair of off-zones is calculated. This difference was compared with the statistical uncertainty and no excess was seen beyond the expected statistical fluctuations. Secondly, the number of events in the on-zone together with the average number of events in the three off-zones is tested using the simulated atmospheric background and the difference was found to be within the expectation from the statistical uncertainty. The details of this procedure can be found in the following section.

5.3 Check for the Number of Background Events in the On-Zone and the Off-Zones

Difference in the number of background events between the on-zone and the three off-zones was tested. First, the average number of events in the three off-zones, $\overline{n}$, was plotted versus the following value:

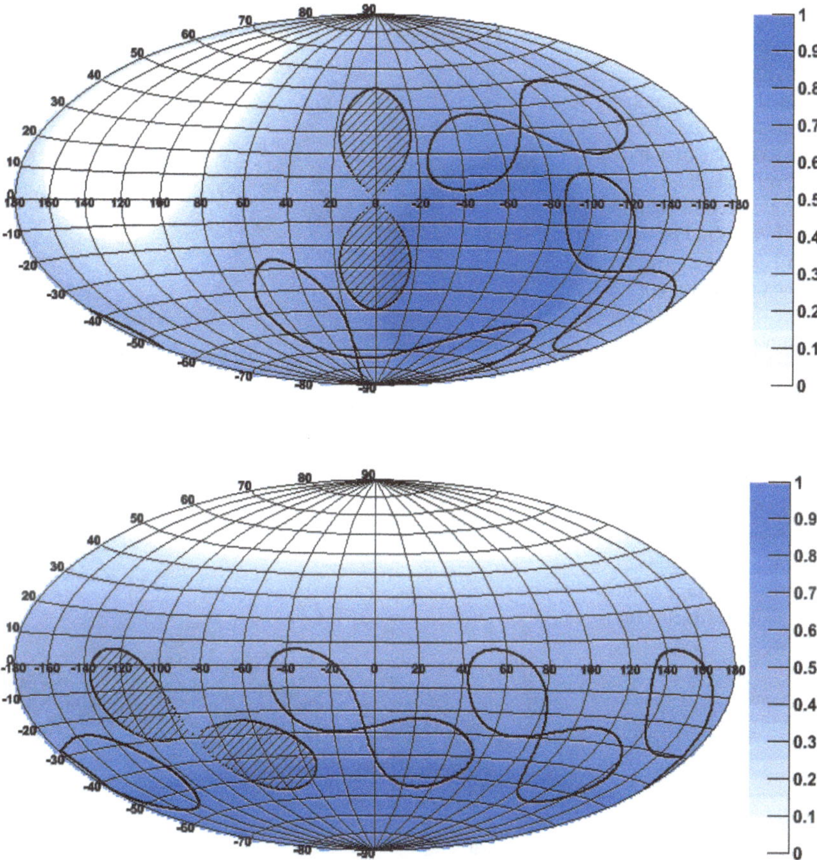

Fig. 5.3 Hammer equal-area map projection (Sect. 4.2.2) in galactic (*top*) and equatorial (*bottom*) coordinates showing the Fermi bubble regions (on-zone) *shaded area* in the centre. The Fermi bubbles area (on-zone) is the shaded one in the centre. The regions corresponding to the three off-zones are also depicted. The colour fill represents the visibility of the sky at the ANTARES site. The maximum on the colour scale corresponds to a 24 h per day visibility

$$\text{difference}_{i,j} = \frac{n_i - n_j}{\overline{n}} \times 100\,\%, \tag{5.6}$$

where n_i is the number of events in the ith off-zone with the cut (Λ^{cut}, $E_{\text{Rec}}^{\text{cut}}$). This plot is shown in Fig. 5.4. The grey area represents the statistical uncertainty estimated for this value. Points have the following colours: black for the difference between the first and the second off-zone, red for the second and the third and green for the third and the first off-zones. The off-zones are numbered from the right to left in equatorial coordinates (Fig. 5.3). About 68 % of the points are expected to lay inside the grey area in the case of same visibility of the zones. It can be seen that most of the points are inside the grey area, even if black points are systematically out for the

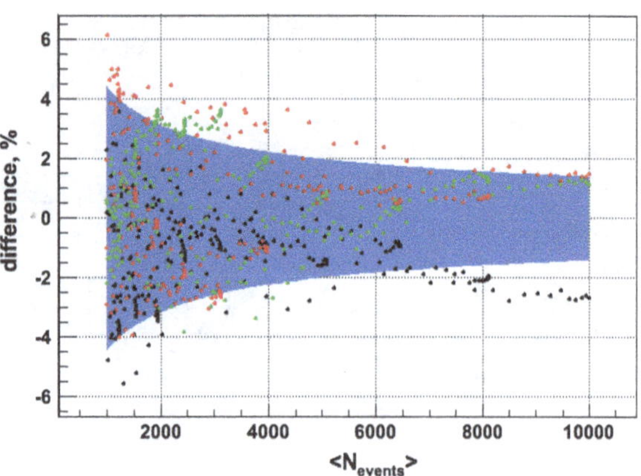

Fig. 5.4 Difference of the number of events between the off-zones expressed as a fraction of average number of events for different cuts (Λ, E_{Rec}) for the first and the second off-zone (*black points*), the second and the third off-zone (*red points*), the third and the first off-zone (*green points*) together with the statistical uncertainty (*grey area*)

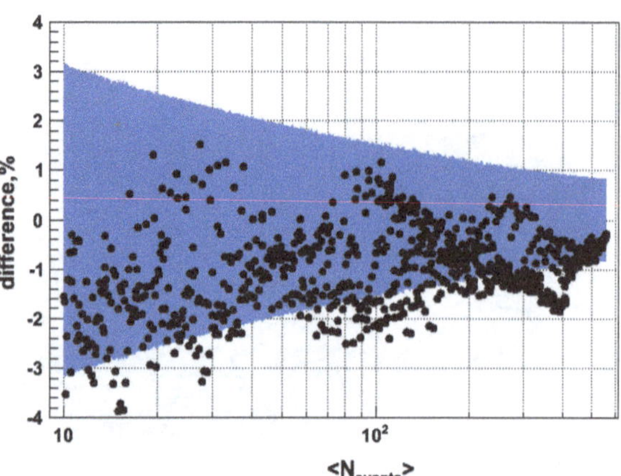

Fig. 5.5 Difference of the number of events between the on-zones and the average of the off-zones expressed as a fraction of average number of events for different cuts (Λ^{cut}, $E^{\text{cut}}_{\text{Rec}}$)

right side of the plot. This systematic behaviour can be ascribed to the procedure which, actually, does not create a set of independent measurements. Indeed, for the two similar cuts ($\Lambda^{\text{cut}}_1 \approx \Lambda^{\text{cut}}_2$ and $E^{\text{cut}}_1 \approx E^{\text{cut}}_2$) the selected events are almost the same. So, if for one selection the ratio is out from the 68 % band (even due to the statistical fluctuation), this will most probably happen for the similar cuts (which correspond to the nearby points on the plot).

The similar check was done for the on-zone using simulated atmospheric neutrinos. In Fig. 5.5 $\overline{n}_{\text{off-zone}}$ is plotted versus the following quantity:

$$\text{difference} = \frac{n_{\text{on-zone}} - \overline{n}_{\text{off-zone}}}{\overline{n}_{\text{off-zone}}} \times 100\,\%. \tag{5.7}$$

Similarly, most of the events are in the grey area. Another representation of this analysis can be done as following. Assuming that for every cut the distribution of the plotted difference has a Gaussian distribution, one can plot the normalised difference:

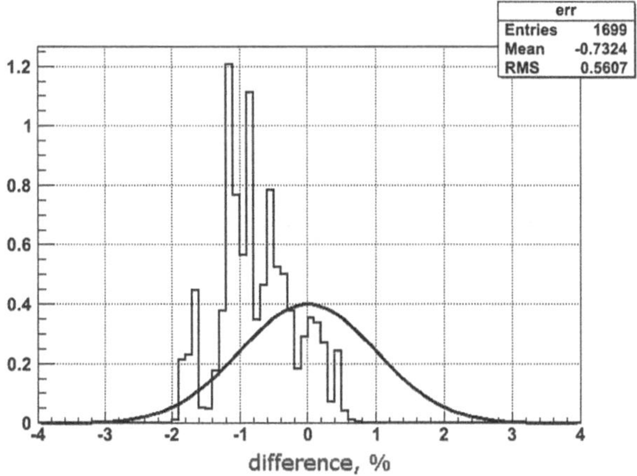

Fig. 5.6 Difference of the number of events between the on-zones and the average of the off-zones expressed as a fraction of average number of events and normalised on the statistical uncertainty for different cuts (Λ^{cut}, $E_{\text{Rec}}^{\text{cut}}$). The smooth line is a Gaussian distribution with mean 0 and sigma 1

$$\text{difference}_{\text{norm}} = \frac{\frac{n_{\text{on-zone}} - \overline{n}_{\text{off-zone}}}{\overline{n}_{\text{off-zone}}} \times 100\,\%}{\text{stat. error.}}, \tag{5.8}$$

which should be a Gaussian distribution with mean 0 and variance 1 in case of the same expected number of events in the on-zone and the off-zones. The distribution of the normalised difference together with a Gaussian is shown in Fig. 5.6. It can be seen that the experimental distribution is slightly different from the Gaussian function. Although the estimated deviation of the mean value is on the level of 1 %. This difference, again, can be ascribed to the intrinsic bias in the measurements (for the two similar cuts the event sets are almost the same).

As a conclusion, there is no obvious reason to consider the presence of the systematic differences between the zones in the current analysis for any event selection. The difference seen is ascribed to the statistical uncertainties only. The measurements in the proposed approach can not be considered as independent. Finally, the systematic uncertainty which can be guessed from the figures presented is at maximum 1 % which does not alter significantly the final result of this work (an upper limit).

5.4 Event Selection Criteria

The analysis adopts a blind strategy in which the cut optimisation is performed using simulated data for the signal and the background. The main quantities used to discriminate between the cosmic neutrino candidate events and the background from misreconstructed atmospheric muons and from atmospheric neutrinos are the

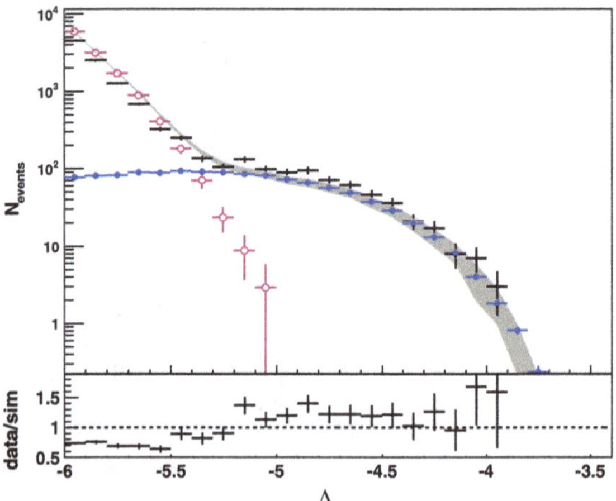

Fig. 5.7 Distribution of the fit-quality parameter Λ for the upgoing events arriving from the three off-zones: data (*black crosses*), 68 % confidence area given by the total background simulation (*grey area*), $\nu_{\rm atm}^{\rm sim}$ (*blue filled circles*), $\mu_{\rm atm}^{\rm sim}$ (*pink empty circles*); bin-ratio of the data to the total background simulation (*bottom*)

tracking quality parameter Λ and the reconstructed energy $E_{\rm Rec}$. The reconstructed energy $E_{\rm Rec}$ is used to reject the atmospheric neutrino background while Λ is used mostly to reject atmospheric muons.

Data in the period from May 2008, when the detector started to operate in its complete configuration, until December 2011 are used. In total 806 days are selected for the analysis. The data runs selection is described in details in Sect. 4.3.2. Figure 5.7 shows the distribution of data and simulated events as a function of the parameter Λ for events arriving from the three off-zones. Here events with at least 10 detected photons and an angular error estimate $\beta < 1°$ are selected. The requirement on the number of photons removes most of the low-energy background events. The angular error condition is necessary in order to ensure a high angular resolution to avoid events originating from an off-zone region being associated with the signal region and vice versa.

At $\Lambda \sim -5.3$ the main background component changes from the misreconstructed atmospheric muons to the upgoing atmospheric neutrino events as seen in Fig. 5.7. The flux of atmospheric neutrinos in the simulation is 23 % lower than observed in the data. This is well within the systematic uncertainty on the atmospheric neutrino flux and the flux from the simulations was scaled accordingly in the following analysis.

A comparison of the energy estimator for data and for atmospheric neutrino simulation is shown in Fig. 5.8 for the same event selection but with a stricter cut $\Lambda > -5.1$ to remove most of the misreconstructed atmospheric muons. The reconstructed energy of the simulated events has been shifted, $\log_{10} E_{\rm Rec}^{\rm shifted} = \log_{10} E_{\rm Rec} + 0.1$, in order to improve the agreement between data and simulations. This is within the estimated uncertainty of the optical module efficiency and the water absorption length (Palioselitis 2012).

This study was done using two simulated background samples. The first sample uses older angular acceptance and standard absorption length curve (Fig. 3.5). The second sample uses a more recent and wider angular acceptance parametrisation

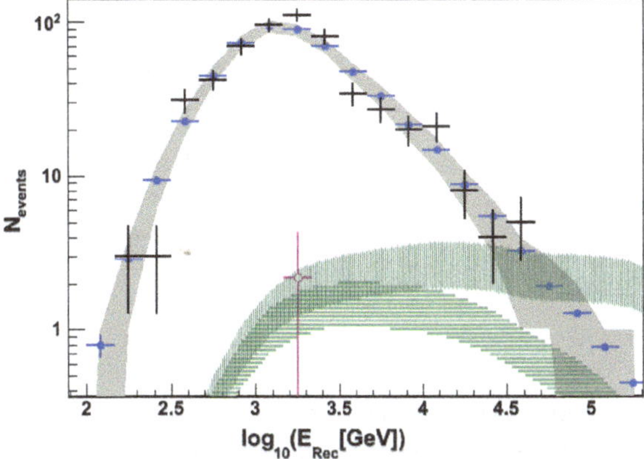

Fig. 5.8 E_{Rec} distribution of the events arriving from the three off-zones with $\Lambda > -5.1$: data (*black crosses*), 68 % confidence area for the total background from simulation (*grey area*), $\nu_{\mathrm{atm}}^{\mathrm{sim}}$ (*blue filled circles*), $\mu_{\mathrm{atm}}^{\mathrm{sim}}$ (*pink empty circles*), expected signal from the Fermi bubbles without neutrino-energy cutoff (*green area filled with dots*) and 50 TeV cutoff (*green area filled with horizontal lines*)

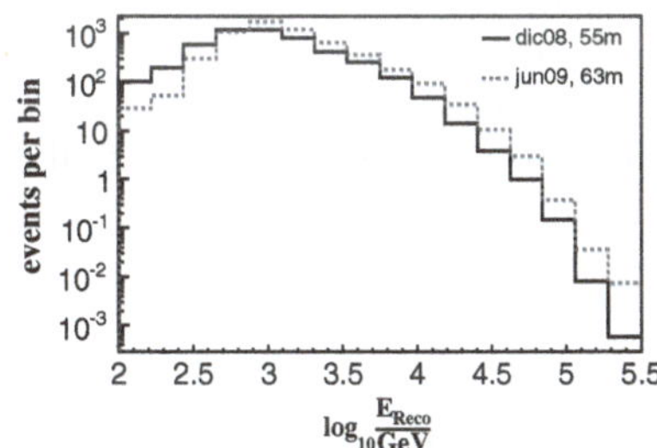

Fig. 5.9 Distribution of the reconstructed energies for simulation with older OM angular acceptance and measured light absorption length in water (*solid*) compared to simulation with newer angular acceptance parametrisation and rescaled absorption length (*dotted*). This image is taken from (Palioselitis 2012)

and absorption length with a maximum of 63 m. The only difference between the two assumptions for the absorption length is a scale factor. The main effect should come from the absorption lengths. Figure 5.9 shows the difference between the reconstructed energy distributions for the two different simulations. The distribution that corresponds to a longer absorption length appears shifted towards higher energies at order of 0.1 in logarithmic scale. This shift can be understood as follows. A muon with a certain fixed energy will give more light in the OM's when the assumed absorption length is longer. Therefore the reconstructed spectrum will be depleted of lower energy entries and exhibit an abundance at higher energies.

Another source of uncertainty is the relative OM efficiency. A lower OM efficiency than what is obtained in calibration can have a large effect on the event rates and energy distribution. The actual efficiency is not expected to be lower than 90 % of the expected one. To test the effect, additional simulation sample was done with a

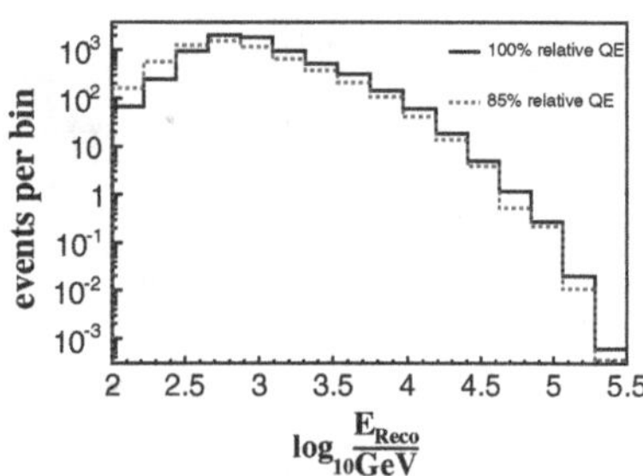

Fig. 5.10 Distribution of the reconstructed energies for a simulation with measured OM efficiency (*solid*) compared to a simulation with reduced OM efficiency (*dotted*). This image is taken from (Palioselitis 2012)

reduced OM efficiency by 15 %. The reduction in the efficiency is applied at the trigger level and it affects the number of triggered as well as reconstructed events. The distributions of the reconstructed energy is shown in Fig. 5.10. As expected, when one overestimates the efficiency of the modules and consequently the amount of light that is detected, the expected distribution is shifted towards higher energies, appearing more rich in high energies than what it is in reality. The shift order is again at level of 0.1 in logarithmic scale. A similar study was performed in this work for the simulated signal events varying OM efficiency and absorption length by $\pm 10\,\%$. This systematic uncertainty study is important for the flux upper limits calculation and it will be discussed in Sect. 5.5.2.

The final event selection is optimised by minimising the average upper limit on the flux:

$$\overline{\Phi}_{90\%} = \Phi_{\nu_\mu+\overline{\nu}_\mu} \frac{\overline{s}_{90\%}(b)}{s}, \tag{5.9}$$

where s is the number of events simulated with the flux $\Phi_{\nu_\mu+\overline{\nu}_\mu}$ from (5.3). The method uses the unified approach of Feldman and Cousins (1998) to calculate signal upper limits with 90 % confidence level, $\overline{s}_{90\%}(b)$, for a known number of background events b. This best average upper limit in the case of no discovery represents the sensitivity of the detector to the Fermi bubbles' flux (Hill and Rawlins 2003). Using (5.3) the average upper limit on the flux coefficient A can be defined as:

$$\overline{A}_{90\%} = A_{\text{theory}} \frac{\overline{s}_{90\%}(b)}{s}. \tag{5.10}$$

Table 5.1 reports the optimal cuts (Λ^{cut}, $E_{\text{Rec}}^{\text{cut}}$) obtained for the four chosen cutoff energies (∞, 500, 100, 50 TeV) of the neutrino source spectrum and the corresponding value of the average upper limit on the flux coefficient $\overline{A}_{90\%}$. Additionally, the optimal cuts for $E_\nu^{\text{cutoff}} = 100$ TeV are applied for the other neutrino-energy cutoffs and the values $\overline{A}_{90\%}^{100}$ are reported for comparison. As the obtained values $\overline{A}_{90\%}$ and $\overline{A}_{90\%}^{100}$ for each cutoff are similar, the 100 TeV cuts are chosen for the final event selection.

The cut optimisation was done using only conventional atmospheric background component (which is due to pion and kaon decays). The optimal event selection

Table 5.1 Optimisation results for each cutoff of the neutrino energy spectrum

E_{ν}^{cutoff} (TeV)	∞	500	100	50
Λ^{cut}	−5.16	−5.14	−5.14*	−5.14
$\log_{10}(E_{\text{Rec}}^{\text{cut}}[\text{GeV}])$	4.57	4.27	4.03*	3.87
$\overline{A}_{90\%}$	2.67	4.47	8.44	12.43
$\overline{A}_{90\%}^{100}$ (100 TeV cuts)	3.07	4.68	8.44	12.75

Average upper limits on the flux coefficient $\overline{A}_{90\%}$ are presented in units of 10^{-7} GeV cm^{-2} s^{-1}sr^{-1}. Numbers with a star indicate the cut used for the $\overline{A}_{90\%}^{100}$ calculation presented in the last row of the table

cuts (Table 5.1) are ~10 TeV. Recent calculations of prompt atmospheric component predict a negligible contribution to the atmospheric neutrino flux below these cuts (Fig. 3.2, Enberg et al. 2008; Martin et al. 2003). This ensures that the obtained cuts does not depend significantly from the prompt component simulation. Also the final analysis is based on the comparison of on- and off-zones (Sect. 5.2), so atmospheric background simulation is not involved in calculation of significance and upper limits described in the following section.

5.5 Results

The final event selection with $\Lambda^{\text{cut}} = -5.14$, $\log_{10}(E_{\text{Rec}}^{\text{cut}}[\text{GeV}]) = 4.03$ is applied to the unblinded data. In the three off-zones 9, 12 and 12 events are observed. In the Fermi bubble regions $n_{\text{obs}} = 16$ events are measured. The events in the Fermi bubbles regions are shown in Fig. 5.11. No events clustering is observed.

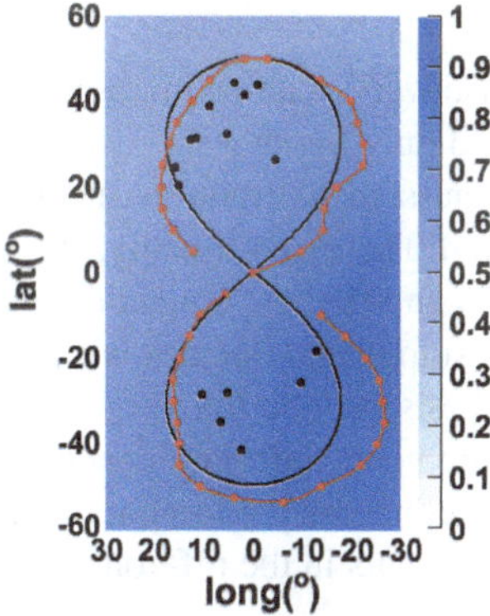

Fig. 5.11 Events in the Fermi bubble regions after the final cut (*black circles*) together with the approximate edges (Su et al. 2010) (*red line*, *circles*) and the simplified shape used in this analysis (*black line*) in galactic coordinates. The colour fill represents the visibility of the sky at the ANTARES site. The maximum on the colour scale corresponds to a 24 h per day visibility

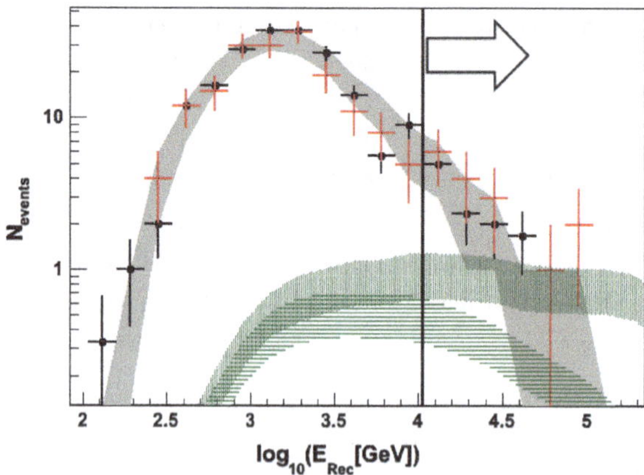

Fig. 5.12 Distribution of the reconstructed energy of the events after the final cut on Λ: events in on-zone (*red crosses*), average over off-zones (*black circles*), 68 % confidence area given by the total background simulation (*grey area*), expected signal from the Fermi bubbles without neutrino-energy cutoff (*green area filled with dots*) and 50 TeV cutoff (*green area filled with horizontal lines*). The chosen $E_{\text{Rec}}^{\text{cut}}$ is represented by the *black line* with an *arrow*

The distribution of the energy estimator for both the on-zone and the average of the off-zones is presented in Fig. 5.12. A small excess of high-energy events in the on-zone is seen with respect to both the average from the off-zones and the atmospheric neutrino simulation. In the following sections the significance estimation and upper limits calculation are described.

5.5.1 Significance Estimation

The significance of signal discovery is an estimation how incompatible is the measurement with the background only hypothesis. Let us suppose that a process has a background described with Gaussian distribution (mean b and a sigma σ). Let us also suppose that n_{obs} was measured. Then, traditionally, the significance can be estimated as $S = (n_{\text{obs}} - b)/\sigma$ and one can say that "S standard deviation was observed". For more complicated processes some test statistics which has a Gaussian distribution for a background only hypothesis may be found.

For the on/off-zone measurements the significance calculation is presented in (Li and Ma 1983). For such measurements the total number of events in the on-zone is compared with the number of events in the off-zone using the following probability density function:

$$\mathcal{L}(n_{\text{obs}}, n_{\text{bg}} \mid s, b; \tau) = \text{Poisson}(n_{\text{obs}} \mid b + s)\text{Poisson}(n_{\text{bg}} \mid \tau b), \qquad (5.11)$$

where s is an unknown signal, b is a background mean, τ is a ratio between the off-zone and the on-zone total areas, n_{bg} is a number of events measured in the total off-zone area. Significance is calculated using the following likelihood ratio test:

$$\lambda = \frac{\mathcal{L}(n_{\text{obs}}, n_{\text{bg}} \mid s = 0, \hat{b})}{\mathcal{L}(n_{\text{obs}}, n_{\text{bg}} \mid \hat{s}, \hat{b})}, \tag{5.12}$$

where the likelihood maximum for the no signal hypothesis is found for the numerator and the best likelihood for all hypotheses is found for the denominator. The value of $-2\ln\lambda$ has approximately χ^2 distribution for the no signal true model (Casella and Berger 2001, Sect. 10.3.1). Therefore, $\sqrt{-2\ln\lambda}$ has approximately a Gaussian distribution and may be used as a significance estimation. The significance equation becomes:

$$S = \sqrt{2}\left\{n_{\text{obs}}\ln\left[\frac{1+\alpha}{\alpha}\left(\frac{n_{\text{obs}}}{n_{\text{obs}}+n_{\text{bg}}}\right)\right] + n_{\text{bg}}\ln\left[(1+\alpha)\left(\frac{n_{\text{bg}}}{n_{\text{obs}}+n_{\text{bg}}}\right)\right]\right\}^{1/2}, \tag{5.13}$$

where $\alpha = 1/\tau$. This significance calculation was tested with the Toy Monte Carlo simulation of the experiment with no signal and background $b = 11$ per zone. For each Toy MC significance was calculated using (5.13) and added to histogram. This histogram is presented in Fig. 5.13 together with Gaussian distribution with mean 0 and sigma 1. One can see that the calculated values are well approximated with Gaussian distribution. This means that the formula (5.13) presents valid significance estimation.

A significance 1.2 σ as a standard deviation of the no-signal hypothesis was obtained from the unblinded data using (5.13).

An extended likelihood is introduced to describe also the event energy distribution:

$$\mathcal{L}(n_{\text{obs}}, n_{\text{bg}} \mid s, b; \tau) = \text{Poisson}(n_{\text{obs}} \mid b + s)\text{Poisson}(n_{\text{bg}} \mid \tau b)\prod_{i=1}^{n_{\text{obs}}} p(E_i \mid s, b), \tag{5.14}$$

This likelihood should provide better estimation of the significance since it uses more information than (5.11). Probability density function $p(E_i \mid s, b)$ can be obtained from the signal and background simulation:

$$h(E \mid s, b) = s\frac{h_s(E)}{\int_{\text{cut}}^{\infty} h_s(E)\,\mathrm{d}x} + b\frac{h_b(E)}{\int_{\text{cut}}^{\infty} h_b(E)\,\mathrm{d}x}, \tag{5.15}$$

which is then normalised to 1:

$$p(E \mid s, b) = \frac{h(E \mid s, b)}{\int_0^{\infty} h(E \mid s, b)\,\mathrm{d}x}. \tag{5.16}$$

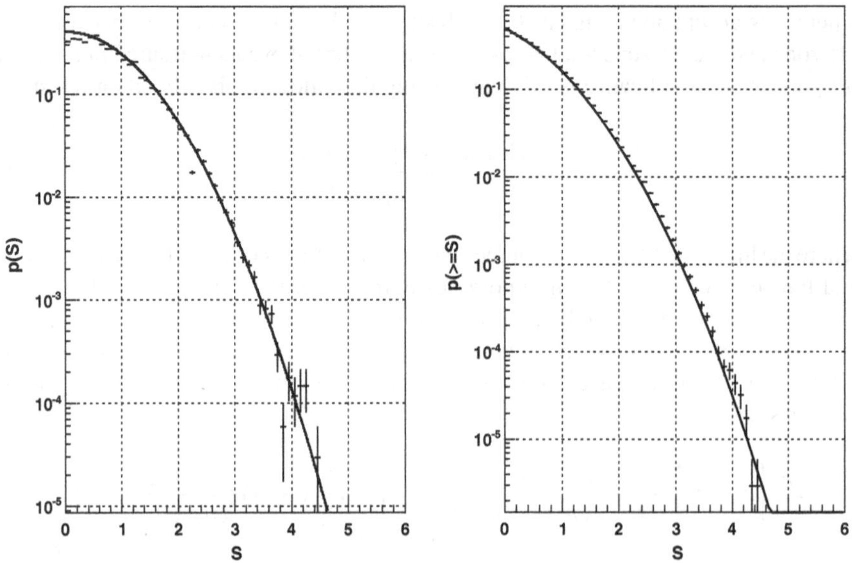

Fig. 5.13 Distribution (*left*) and cumulative distribution (*right*) of the significance calculated with Li and Ma method for the Toy Monte Carlo simulations of the on/off-zone measurements with no signal, mean background $b = 11$ and $\tau = 3$ (*black*), compared with a Gaussian distribution with mean 0 and sigma 1

Signal simulation without energy cutoff was used for this calculation. To implement significance calculations in the ROOT framework, histograms were used for $h_s(E)$, $h_b(E)$ and $p(E_i \mid s, b)$. This is less precise than using fitted functions, but for a quite fine binning the difference should be negligible. Significance estimation follows the previously described method except the fact that $-2\ln\lambda$ can not be approximated with χ^2 distribution.[1] Toy Monte Carlo simulation (Toy MC) was used to estimate the significance. Each Toy MC simulates the experiment with no true signal. To do such simulations in the on-zone, firstly, average total number of the events is calculated integrating $h_b(E)$ distribution, which rescaled to have 11 events after the cut. After, for each Toy the number of events is randomly generated with a Poisson distribution using the calculated average value. Lately, the event energies are sampled from the background distribution. For the off-zones the procedure is the same, except, that for the first step $h_b(E)$ is rescaled to have 33 events after the cut. For the simulated event energy distributions $\sqrt{-2\ln\lambda}$ is calculated according to (5.14).

For the event energy distributions from the data $\sqrt{-2\ln\lambda} = 0.86$ is obtained. Simulation shows that 81 % of Toy MCs have $\sqrt{-2\ln\lambda} < 0.86$ (1000 Toy MCs were used). This p-value corresponds to 1.3 σ significance of the signal discovery.

Examples of the obtained event distributions in the on-zone are shown in Fig. 5.14. As it can be seen, there are events with $\log_{10} E_{\mathrm{Rec}} > 4.7$ almost in every histogram.

[1] This was found by running Toy Monte Carlo simulation.

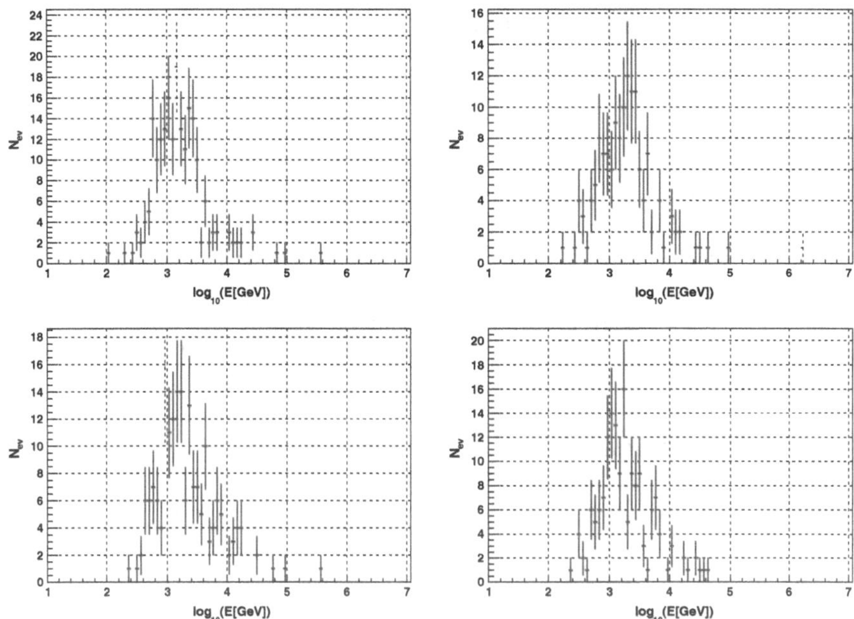

Fig. 5.14 Toy MC event-energy distributions

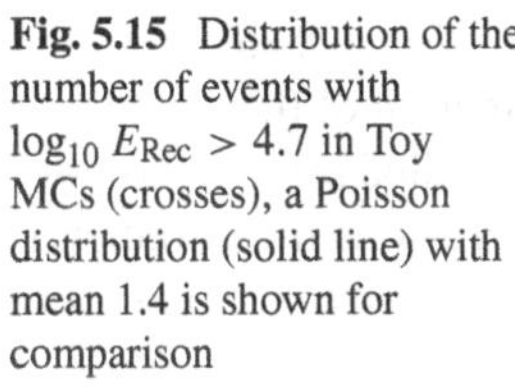

Fig. 5.15 Distribution of the number of events with $\log_{10} E_{\mathrm{Rec}} > 4.7$ in Toy MCs (crosses), a Poisson distribution (solid line) with mean 1.4 is shown for comparison

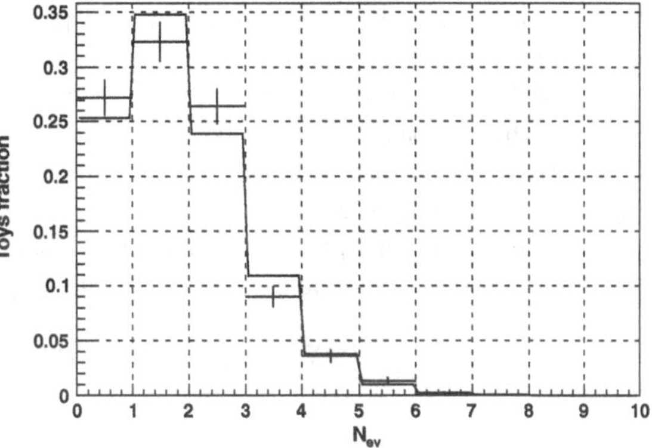

The probability for this is actually not so low as it seems from the first look. Integration of the simulated background with $\log_{10} E_{\mathrm{Rec}} > 4.7$ estimates 1.4 events in average. Also, distribution of the number of events with $\log_{10} E_{\mathrm{Rec}} > 4.7$ in Toy MCs fits quite well a Poisson distribution with mean 1.4 as it is shown in Fig. 5.15.

As a conclusion, the obtained 1.3 σ significance using the event energy distribution is very similar to 1.2 σ significance calculated with Li & Ma method. The latter method is simpler and better known, so for the journal publication it was decided to use it as official significance calculation.

5.5.2 Upper Limits Calculation

The main goal of any signal measurement is to estimate the signal value and the precision of the estimation. For this purpose the estimated signal range are provided as a result of the measurement. The probability of the signal to be between the range limits is fixed to some value (90 %, for example, or $1 - \alpha$). In terms of statistics this means that the measurement X was done and one wants to provide limits $s_{\min}$, $s_{\max}$ for which $P(s_{\min} < s < s_{\max} \mid X) = 1 - \alpha$. The values $s_{\min}$, $s_{\max}$ can be calculated if $P(s \mid X)$ is known for every s. Two different types approaches to define parameter ranges were developed: Bayesian and frequentists. They will be described in following sections and results are compared after.

5.5.2.1 Bayesian Approach

In most cases it is straightforward to know the probability $P(X \mid s)$ of observing X if the signal s is known. This can be done assuming Poisson or another model involved in the measurements. To obtain $P(s \mid X)$ one can use the Bayes' theorem:

$$P(s \mid X) = \frac{P(X \mid s)P(s)}{P(X)}, \tag{5.17}$$

where the denominator is simply a constant as it depends only on the measurement X, so it can be calculated using the probability property $\int_{-\infty}^{+\infty} P(s \mid X)\,\mathrm{d}s = 1$. The remaining multiplier $P(s)$ is the unknown probability distribution of the signal.

To find $P(s \mid X)$ from the formula (5.17) one can use a so-called prior probability $P_{\text{prior}}(s)$ which is based on some belief, physical limits or previous measurements. This approach is called Bayesian. $P(s \mid X)$ obtained from the formula (5.17) using $P_{\text{prior}}(s)$ is named a posterior probability.

In case of several independent parameters (signal with background, for example), probability properties can be used together with Bayes' theorem to obtain:

$$P(s, b \mid X) = \frac{P(X \mid s, b)P(s, b)}{P(X)} = \frac{P(X \mid s, b)P(s)P(b)}{P(X)}. \tag{5.18}$$

Parameters such as background are normally out of interest. These nuisance parameters can be excluded from the posterior probability by the following marginalisation procedure:

$$P(s \mid X) = \int_{-\infty}^{\infty} P(s, b \mid X)\,\mathrm{d}b. \tag{5.19}$$

Marginalisation procedure is a quantitative way to say "I don't know, and I don't care" for the nuisance parameters (Loredo 1990). It is worth to mention, that marginalisation favours regions of parameter space that contain a large volume of the probability density in the marginalised directions. This "volume effect" can sometimes lead to counter-intuitive results, such as suppression of the probability density for the global best fit parameters if they appear within sharp peaks or ridges that contain little volume (Hamann et al. 2007). So, it is dangerous to use marginalised post probabilities for the best value estimation (although in the case of the Gaussian distributions result will be correct). Additionally, the volume effect changes with different priors, so for the reliability test it is recommended to check several priors (or parameter choices, for example, $\ln s$ instead of s).

Bayesian approach for the limits estimation may be summarised as following:

- Construct probability function for the measurement $P(X \mid s, b)$ using a knowledge of the involved processes (Poisson, exponential, etc.). If it is considered as a function of s, b it is also called likelihood $\mathcal{L}(s, b \mid X)$.
- Estimate prior probabilities for all parameters.
- Marginalise posterior probability to exclude nuisance parameters.
- Find interval $(s_{\min}, s_{\max})$ for which $P(s_{\min} < s < s_{\max} \mid X) = 1 - \alpha$. This interval can be chosen unambiguously if values of s are included to the interval by decreasing order of $P(s \mid X)$. The interval found in this way is called minimum credible interval (in case of several local maximums it can be a set of intervals).

To construct the probability function, first, let us describe the measurement:

One measurement of the background is done in the absence of the signal. The obtained value is 33 events. Another measurement is done in the area where the signal may be present. The obtained value is 16. The area for the signal measurement is 3 times smaller than the area for the background measurements. The signal flux may be estimated from a simulation. The simulation provides the number of events for a known flux. Also the simulation has a known systematic error on the number of events estimation.

In order to account for systematic uncertainties in the simulation of the signal, a dedicated study has been performed in which the assumed absorption length in seawater is varied by ±10 % and the assumed optical module efficiency is varied by ±10 %. For each variation the number of events is calculated for each cutoff and compared with the number of signal events s obtained using the standard simulation. The differences are calculated and summed in quadrature to obtain σ_{syst}.

First what can be noticed is that the measured number of events is quite low (16, 33). For such measurements of counting number of events with a low appearance probability during a fixed time and space Poisson distributions are used. So, assuming an unknown signal s, background b in the on-zone, the two measurements can be described as:

$$\mathcal{L}(n_{\text{obs}}, n_{\text{bg}} \mid s, b) = \text{Poisson}(n_{\text{obs}} \mid b + s) \times \text{Poisson}(n_{\text{bg}} \mid \tau b), \tag{5.20}$$

where $\tau = 3$ shows that the area for the background measurement is three times larger than the area for the signal measurement.

Simulation provides conversion of the flux to the number of the signal events using a simple proportion:

$$\frac{s}{f} = \frac{s_{\text{sim}}}{f_{\text{sim}}}. \tag{5.21}$$

Simulation gives a number of signal events which is affected by a systematic error. One can assume a Gaussian distribution for the simulated value. This Gaussian has and unknown mean which is a true signal θ_{sim} and a known sigma σ_{syst}:

$$P(s_{\text{sim}} \mid \theta_{\text{sim}}) = \text{Gaussian}(s_{\text{sim}} \mid \theta_{\text{sim}}, \sigma_{\text{syst}}). \tag{5.22}$$

So, for the likelihood one can summarise:

$$\mathcal{L}(n_{\text{obs}}, n_{\text{bg}}, s_{\text{sim}} \mid f, b, \theta_{\text{sim}}) = \text{Poisson}(n_{\text{obs}} \mid b + f\frac{s'_{\text{sim}}}{f_{\text{sim}}}) \times \text{Poisson}(n_{\text{bg}} \mid \tau b) \\ \times \text{Gaussian}(s_{\text{sim}} \mid \theta_{\text{sim}}, \sigma_{\text{syst}}). \tag{5.23}$$

The priors for f, b and θ_{sim} can be flat in the range ≥ 0. These priors are improper in the sense that their integral is not 1. However, this is allowed since the post probability can be renormalised in the end.

So, the post probability becomes:

$$P(f \mid n_{\text{obs}}, n_{\text{bg}}, s_{\text{sim}}) \propto \int_0^{\infty}\int_0^{\infty} \text{Poisson}(n_{\text{obs}} \mid b + f\frac{s_{\text{sim}}}{f_{\text{sim}}}) \times \text{Poisson}(n_{\text{bg}} \mid \tau b) \\ \times \text{Gaussian}(s_{\text{sim}} \mid \theta_{\text{sim}}, \sigma_{\text{syst}})\, \mathrm{d}b\, \mathrm{d}\theta_{\text{sim}}, \tag{5.24}$$

which is needed to be normalised to satisfy the following property:

$$\int_0^{\infty} P(f \mid n_{\text{obs}}, n_{\text{bg}}, s_{\text{sim}})\, \mathrm{d}f = 1. \tag{5.25}$$

Additionally, the so-called least informative priors were tested. In case of the Poisson process where the mean λ is measured, it is suggested to use $P(\lambda) = 1/\lambda$ as the least informative prior. This prior is also known as "Jeffreys' prior" (Jeffreys 1939; Loredo 1990). Which becomes for the flux and background the following:

$$P(b) = \frac{1}{b}, \tag{5.26}$$

$$P(b) = \frac{1}{b + f\frac{s_{\text{sim}}}{f_{\text{sim}}}}. \tag{5.27}$$

As the final limits obtained with the marginalisation procedure depend from the prior choice, a complementary approach was used to test the limits. Instead of marginalisation, a profile posterior can used, which is defined as following:

$$\hat{P}(f \mid n_{\text{obs}}, n_{\text{bg}}, s_{\text{sim}}) \propto \max_{b,\theta_{\text{sim}}} P(f, b, \theta_{\text{sim}} \mid n_{\text{obs}}, n_{\text{bg}}, s_{\text{sim}}). \tag{5.28}$$

This function preserves the best value which maximises original post probability (5.24). Also the best value is independent on the prior probabilities choice if the information in likelihood overwhelms the information in priors. The profile likelihood does not have a formal probabilistic interpretation, however it may be used to construct approximate frequentist confidence intervals based on the likelihood ratio test statistic:

$$\chi(f) = \sqrt{-2 \ln \frac{\hat{P}(f)}{\hat{P}(\hat{f})}}, \tag{5.29}$$

where denominator maximises the probability. The test statistic has approximately Gaussian distribution so flux region for which $\chi(f) < 1$ should approximately correspond to 68 % confidence interval (90 % interval can be obtained in similar way). Other frequentist methods which were used in this thesis are described in the next section.

The described post probabilities for the flat and Jeffrey's priors are compared in Fig. 5.16. A slight peak shift and the shape change can be noticed. However, as the change is minimal, the post probability is considered robust to the prior choice. Actually, this means that the measurement itself provides much more information than the priors. Normalised profile likelihood, obtained from the post probability with flat priors (5.24) is shown in the same figure. As it can be seen, its maximum is, again, slightly different from the maximums of the marginalised post probabilities and it has similar shape. In this work, as additional limits test, the profile likelihood was interpreted as a post probability and Bayesian limits were calculated with it.

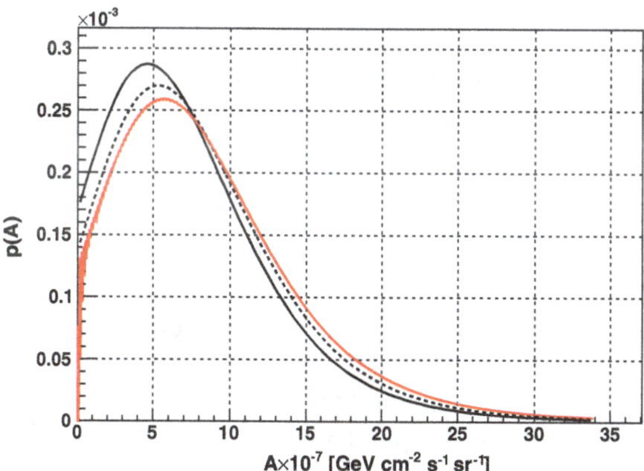

Fig. 5.16 Marginalised post probabilities for Jeffrey's priors (*black solid*) and flat priors (*black dashed*) compared with the normalised profile likelihood (*red solid*) for simulated signal with 100 TeV energy cutoff

5.5.2.2 Frequentist Approach

In contrary to the Bayesian approach, frequentist approach attempts to set limits for repeating experiments. The confidence level $1 - \alpha$ for this approach means that if experiment is repeated unlimited number of times the fraction of experiments where the true value is inside the limits is $1 - \alpha$.

A frequentist method developed by Neyman consists of the two steps (Neyman 1937):

- To find some "confidence belt" for every signal s. The confidence belt is a set of different measurements $\{X^s_n\}$ which give $\sum_{i=1}^{n} \mathcal{L}(s \mid X^s_i) \geq 1 - \alpha$. The way to choose confidence belt $\{X^s_n\}$ is arbitrary.
- To use real measurement X. If X is present in some set of $\{X^s_n\}$ for the signal s then this signal s is included to the confidence area. This operation is done for each set $\{X^s_n\}$ to build the area. The upper limit with $1 - \alpha$ confidence level is the maximum signal in the confidence area.

Limits obtained by this method may vary depending on the choice of the confidence belts. In (Neyman 1937) two options were suggested: central and one-sided limits (upper/lower). Both types have some unwanted features. Central limits may be too strict for the case when the number of the observed events is lower than the expected number from the background. One-sided limits do not allow to claim a discovery or setting the upper limit depending on the measured X. The mentioned problems are essential as one should define which type of the limits one wants to set before the measurement. If this decision is done after the measurement this would lead to the under-coverage of the limits known as a "flip-flopping" problem.

In order to avoid the "flip-flopping" problem the Neyman's method was improved by the special ordering principle (Feldman and Cousins 1998). The advantage is that the procedure to define an upper limit in case of a low number of observed events or a two-sided limit in the opposite case is defined *a priori*. Confidence belt selection is based on sorting the $\{X_n\}$ by the decreasing rank:

$$R = \frac{\mathcal{L}(X \mid s)}{\mathcal{L}(X \mid \hat{s})}, \tag{5.30}$$

where $\mathcal{L}(X \mid \hat{s})$ is a maximum possible likelihood for the fixed measurement X which is realised by the signal $\hat{s}$.

For some cases of the low number of the observed events the measurements with a higher background may provide better limit with this method. Actually, this seeming discomfort is connected with the difference between Bayesian and frequentist approaches. In the latter, if a signal plus background is measured, then if the expected mean background is higher the signal should be lower (in average over a lot of trials). But, on the other hand, it is expected that the experiment with higher expected background can not provide better limits.

For the several parameters this method, however can only provide the area which includes limits on all parameters together. In case of (5.23) it would be $(f, b, \theta_{\mathrm{sim}})$

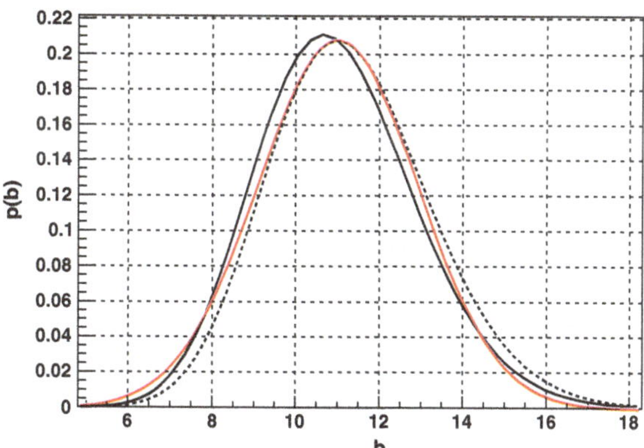

Fig. 5.17 Comparison of a post probability for background b obtained using Jeffrey's prior (*black dashed*) and flat prior (*black solid*) together with a Gaussian approximation (*red solid*)

area. The values of $f_{\min}$, $f_{\max}$ extracted from this appear to be too conservative in comparison with the limits from the previously discussed Bayesian approach. In order to avoid this, semi-Bayesian approach may be used. In this approach likelihood is multiplied by nuisance prior probabilities and marginalised over them. This allows to treat only one parameter (s) in frequentist way. This approach is described in details in Conrad et al. (2003). To use implicitly calculations described in this paper, the post probability for b was approximated with Gaussian distribution with mean $\tilde{b} = (9 + 12 + 12)/3$ and sigma $\sigma_b = \sqrt{9 + 12 + 12}/3$. A comparison of post probability calculated with a Poisson likelihood and Gaussian approximation is shown in Fig. 5.17.

Thus, the likelihood used for the frequentist method becomes:

$$\mathcal{L}(n_{\text{obs}} \mid f) = \int_0^\infty \int_0^\infty \text{Poisson}(n_{\text{obs}} \mid f \frac{s_{\text{sim}}}{f_{\text{sim}}} + b) \times \text{Gaussian}(b \mid \tilde{b}, \sigma_b) \text{Gaussian}(s_{\text{sim}} \mid \theta_{\text{sim}}, \sigma_{\text{syst}})\, \mathrm{d}b\, \mathrm{d}\theta_{\text{sim}}. \tag{5.31}$$

5.5.2.3 Comparison of the Methods

An upper limit on the number of signal events is calculated at 90 % confidence level. The results are summarised in Table 5.2. First, it can be noticed, that flat and Jeffrey's priors in the Bayesian approach provide similar limits. Upper limits with Jeffrey's priors are more stringent which is expected from the comparison of the marginalised post probabilities and the priors (*const* and $1/\lambda$). Limits obtained with the profile likelihood are similar to the limits with flat priors. Also, they are more conservative among the three Bayesian limits. Frequentist upper limits are similar to the Bayesian limits in this table. For the journal paper it was decided to use the upper limits, calculated with the Bayesian approach using profile likelihood.

Table 5.2 90 % confidence level upper limits on the neutrino flux coefficient $A_{90\%}$ for the Fermi bubbles presented in units of 10^{-7} GeV cm^{-2} s^{-1}sr^{-1}

E_{ν}^{cutoff} (TeV)	∞	500	100	50
Number of signal events in simulation s_{sim}	2.9	1.9	1.1	0.7
Uncertainty on the efficiency σ_{syst}, %	14	19	24	27
$A_{90\%}$ with flat priors	5.3	8.4	16.0	24.3
$A_{90\%}$ with Jeffrey's priors	4.9	7.8	15.0	23.0
$A_{90\%}$ with profile likelihood	5.4	8.7	17.0	25.9
$A_{90\%}$ using (Conrad et al. 2003)	5.9	9.1	16.7	26.4

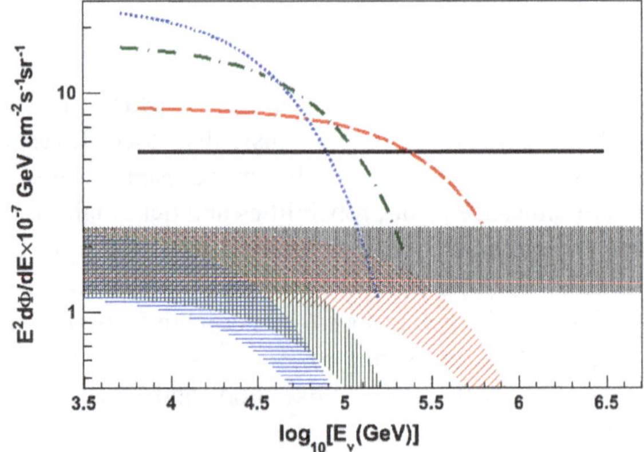

Fig. 5.18 Upper limits on the neutrino flux from the Fermi bubbles for different cutoffs: no cutoff (*black solid*), 500 TeV (*red dashed*), 100 TeV (*green dot-dashed*), 50 TeV (*blue dotted*) together with the theoretical predictions for the case of a purely hadronic model (the same colours, *areas filled with dots*, *inclined lines*, *vertical lines* and *horizontal lines* respectively). The limits are drawn for the energy range where 90 % of the signal is expected

A graphical representation of the upper limits on a possible neutrino flux together with the predicted flux is shown in Fig. 5.18. The limits are drawn for the energy ranges where 90 % of the signal is expected. These ranges are calculated from simulated event energy distributions of a signal with different energy cutoffs. The obtained upper limits are above the expectations from the considered models. Also, the presented limits are worse than it is expected from the average upper limits presented in Table 5.1. This is due to the fact that in this particular measurement a positive excess of events in respect to the background was seen.

The performed analysis presents the first upper limits on high energy neutrino emission from the Fermi bubble regions. Although predictions for neutrino emission are below the ANTARES experiment's sensitivity, the size and location of the ANTARES detector in the Northern Hemisphere suggest that these are the most stringent limits in the high-energy range obtainable today. The sensitivity will improve as more data is accumulated (more than 65 % gain in the sensitivity is expected once 2012–2016 data is added to the analysis).

In the following section a possibility to install a km^3-size detector in the Northern Hemisphere is considered and its sensitivity to the hypothetical neutrino emission from the Fermi bubbles is estimated.

5.6 KM3NeT

The KM3NeT collaboration aims to install a next generation neutrino telescope in the Mediterranean Sea (Aiello et al. 2011). It is expected to have much better sensitivity to the neutrino cosmic sources due to the larger detector volume. The detector sensitivity to the emission from the Fermi bubbles region was estimated by the collaboration (Adrián-Martínez et al. 2013).

The geometry of the detector simulated in this work consists of a three-dimensional array of OMs attached to vertical structures. Each OM consists of 31 3" PMTs (Kavatsyuk et al. 2012) housed inside a 17" pressure-resistant glass sphere covering almost a 4π field of view. In total 12320 OM were distributed in instrumented detector volume of about 6 km^3.

In Fig. 5.19 the discovery fluxes at 5 σ C.L., 50 % probability and 3 σ C.L., 50 % probability for the neutrino spectra considered are shown as functions of the observation time. If the neutrino normalisation factor is of the order of 1×10^{-7} GeV cm^{-2} s^{-1} the discovery of neutrinos from the Fermi bubbles is expected in about one year of data taking for a E^{-2} neutrino spectrum with a cutoff at 100 TeV. The first evidence (3 σ C.L., 50 % probability) could be obtained after a few months of data taking. For the more severe cutoff at 30 TeV the discovery is predicted to be achieved in about 5.5 years and first evidence in about 2.5 years.

Other sensitivity estimations to the Fermi bubbles for the KM3NeT detector may be found in (Cholis 2012; Lunardini and Razzaque 2012).

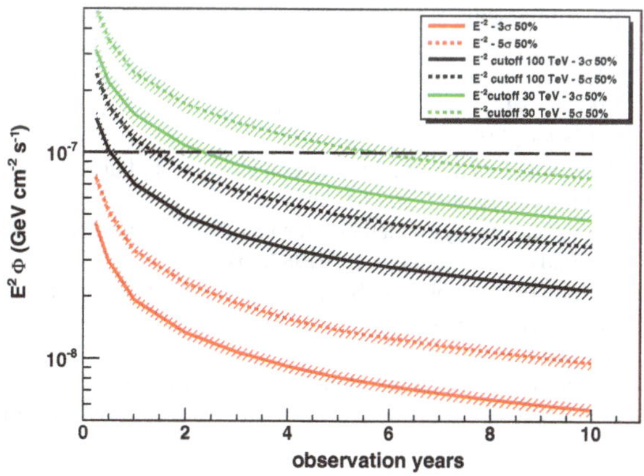

Fig. 5.19 Discovery fluxes as functions of the observation years for 5 σ C.L., 50 % probability and 3 σ C.L., 50 % probability, for the three neutrino spectra assumed. The bands represent the variation due to the uncertainty on the normalisation factor of the conventional Bartol neutrino flux. This image is taken from (Adrián-Martínez et al. 2013)

References

S. Adrián-Martínez et al., Detection potential of the KM3NeT detector for high-energy neutrinos from the Fermi bubbles. Astropart. Phys. **42**, 7–14, (2013). ISSN: 09276505, http://linkinghub.elsevier.com/retrieve/pii/S0927650512002149

S. Aiello et al., KM3NeT technical design report. Technical report. KM3NeT (2011). http://www.km3net.org/TDR/TDRKM3NeT.pdf

E. Carretti et al., Giant magnetized outflows from the centre of the Milky Way. Nature **493**(7430), 66–9 (2013). ISSN: 1476–4687, http://www.ncbi.nlm.nih.gov/pubmed/23282363

G. Casella , R. Berger, *Statistical inference*, 2nd edn. (Duxbury, Pacific Grove, 2001). ISBN: 0-534-24312-6, http://departments.columbian.gwu.edu/statistics/sites/default/files/u20/Syllabus6202-Spring2013-Li.pdf

D. Chernyshov et al., Particle acceleration and the origin of gamma-ray emission from Fermi bubbles. in *Proceedings of the 32nd ICRC*, vol. 1, 2011. http://arxiv.org/abs/1109.2619

I. Cholis, Searching for the High Energy Neutrino counterpart of the Fermi Bubbles signal or from Dark Matter annihilation. 1–14 (2012). http://arxiv.org/abs/1206.1607

J. Conrad et al., Including systematic uncertainties in confidence interval construction for Poisson statistics. Phys. Rev. D **67**(1), 012002 (2003). ISSN: 0556–2821, http://link.aps.org/doi/10.1103/PhysRevD.67.012002

R.M. Crocker, F. Aharonian, Fermi bubbles: giant, multibillion-year-old reservoirs of galactic center cosmic rays. Phys. Lett. B **106**(10), 101102 (2011). ISSN: 0031–9007, http://link.aps.org/doi/10.1103/PhysRevLett.106.101102

G. Dobler, A last look at the microwave haze/bubbles with WMAP. ApJ **750**(1), 17 (2012). ISSN: 0004–637X, http://iopscience.iop.org/0004-637X/750/1/17/

G. Dobler et al., The Fermi gamma-ray haze from dark matter annihilations and anisotropic diffusion. ApJ **741**(1), 25 (2011). ISSN: 0004–637X, url:http://iopscience.iop.org/0004-637X/741/1/25/

R. Enberg et al., Prompt neutrino fluxes from atmospheric charm. Phys. Rev. D **78**(4), 043005 (2008). ISSN: 1550–7998, http://link.aps.org/doi/10.1103/PhysRevD.78.043005

G. J. Feldman, R.D. Cousins, Unified approach to the classical statistical analysis of small signals. Phys. Rev. D **57**(7), 3873–3889 (1998). ISSN: 0556–2821, http://link.aps.org/doi/10.1103/PhysRevD.57.3873

J. Hamann et al., Observational bounds on the cosmic radiation density. JCAP **2007**(08), 21 (2007). http://stacks.iop.org/1475-7516/2007/i=08/a=021

C.G.T. Haslam et al., A 408 MHz all-sky continuum survey. II—the atlas of contour maps. Astron. Astrophys. Suppl. Ser. **47**, 1–142 (1982). http://adsabs.harvard.edu/full/1982A26AS471H

G.C. Hill, K. Rawlins, Unbiased cut selection for optimal upper limits in neutrino detectors: the model rejection potential technique. Astropart. Phys. **19**(3), 393–402 (2003). ISSN: 09276505, http://linkinghub.elsevier.com/retrieve/pii/S0927650502002402

H. Jeffreys, *Theory of Probability* , 3rd edn. (Oxford University Press, Oxford, 1939)

O. Kavatsyuk et al., Multi-PMT optical module for the KM3NeT neutrino telescope. NIM A **695**, 338–341 (2012). ISSN: 01689002, http://linkinghub.elsevier.com/retrieve/pii/S0168900211018584

B.C. Lacki, The Fermi bubbles as starburst wind termination shocks. (2013). http://arxiv.org/abs/1304.6137v1

T. Li, Y. Ma, Analysis methods for results in gamma-ray astronomy. ApJ **272**, 317–324 (1983). http://adsabs.harvard.edu/full/1983ApJ...272.317L7

T. Loredo, From Laplace to Supernova SN 1987A: Bayesian inference in astrophysics. Maximum-entropy and Bayesian meth. 81–142 (1990). http://link.springer.com/chapter/10.1007/978-94-009-0683-9_6

C. Lunardini, S. Razzaque, High energy neutrinos from the Fermi bubbles. Phys. Lett. B **108**(22), 221102 (2012). ISSN 0031–9007, http://link.aps.org/doi/10.1103/PhysRevLett.108.221102

A. Martin et al., Prompt neutrinos from atmospheric cc and bb production and the gluon at very small x. Acta Physica Polonica B **34**, 3273–303 (2003). http://www.actaphys.uj.edu.pl/_cur/store/vol34/pdf/v34p3273.pdf

P. Mertsch, S. Sarkar, Fermi gamma-ray bubbles from stochastic acceleration of electrons. Phys. Lett. B **107**(9), 091101 (2011). ISSN: 0031–9007, http://link.aps.org/doi/10.1103/PhysRevLett.107.091101

J. Neyman, Outline of a theory of statistical estimation based on the classical theory of probability. Philos. Trans. Roy. Soc. A **236**(767), 333–380 (1937). http://www.jstor.org/stable/10.2307/91337

D. Palioselitis, Measurement of the atmospheric neutrino energy spectrum. Ph.D. thesis, Institute for High Energy Physics (IHEF), 2012. http://dare.uva.nl/cgi/arno/show.cgi?fid=444188

S. Snowden, R. Egger, ROSAT survey diffuse X-Ray background maps. II. ApJ **485**, 125–135 (1997). http://iopscience.iop.org/0004-637X/485/1/125

M. Su et al., Giant gamma-ray bubbles from Fermi-Lat: active galactic nucleus activity or bipolar galactic wind? ApJ **724**(2), 1044–1082 (2010). ISSN: 0004–637X, http://iopscience.iop.org/0004-637X/724/2/1044/

S. Thoudam, Fermi bubble γ-rays as a result of diffusive injection of galactic cosmic rays. (2013). http://arxiv.org/abs/1304.6972v1

F. Villante, F. Vissani, How precisely can neutrino emission from supernova remnants be constrained by gamma ray observations? Phys. Rev. D **78**(10), 103007 (2008). ISSN: 1550–7998, http://link.aps.org/doi/10.1103/PhysRevD.78.103007

Conclusion

ANTARES, the first underwater neutrino telescope in the Mediterranean sea, is successfully operating in its final 12-lines configuration since 2008. The INFN group of Genova contributed to the realisation of the detector with the design and construction of various parts.

Before and during my Ph.D. thesis I have participated to both the research and developement phase and to the data collection. In particular I have performed studies of the core detection elements: the photomultipliers. I have also participated to a detailed optical module simulation with GEANT4. I have also worked on the detector calibration, the detector maintenance, the detector operation and the data quality control.

As main scientific analysis, a search for high-energy neutrino emission from the region of the Fermi bubbles has been performed using data from the ANTARES detector. A method for the background estimation using off-zones has been developed specially for this measurement. A new likelihood for the limits calculation which treats both observations in the on-zone and in the off-zone in the similar way and also includes different systematic uncertainties has been constructed.

The analysis of 2008–2011 ANTARES data yielded to a 1.2 σ excess of events in the Fermi bubble regions, compatible with no-signal hypothesis. For the optimistic case of no energy cutoff in the flux, the upper limit is within a factor of three of the prediction of purely hadronic model based on the measured gamma-ray flux. The sensitivity will improve as more data are accumulated (more than 65 % gain in the sensitivity is expected once 2012–2016 data are added to the analysis). Results of this work are published in Adrián-Martínez et al. (2014).

The future KM3NeT telescope is the ideal instrument to observe neutrinos from the Fermi bubbles due its location and large instrumentation volume. In the case of an E_{ν}^{-2} spectrum with an exponential cutoff at 100 TeV, a 3 σ evidence is expected in few months and a discovery after about one year of data taking can be claimed if the purely hadronic model is confirmed. The non-observation of a signal would severely constrain models of neutrino production through hadronic acceleration processes in the Fermi bubbles.

V. Kulikovskiy, *Neutrino Astrophysics with the ANTARES Telescope*,
Springer Theses, DOI 10.1007/978-3-319-20412-3

Reference

S. Adrián-Martínez et al., A search for neutrino emission from the Fermi bubbles with the ANTARES telescope, EPJ C **74**(2), 2701 (2014). ISSN: 1434–6044, http://link.springer.com/10.1140/epjc/s10052-013-2701-6

Curriculam Vitae

Name:	**Vladimir Kulikovskiy**
Date and Place of Birth:	March 27th, 1986, Moscow
E-mail:	Vladimir.Kulikovskiy@ge.infn.it
Citizenship:	Russian Federation

Education:

2002–2008:	Lomonosov Moscow State University, Physics Faculty. Masters degree in nuclear and particle physics, *"Summa Cum Laude"*. Subject of the thesis: "Optimization of the deep underwater neutrino telescope NEMO".
2011–2013:	Università di Genova/Université Paris Diderot (Paris 7). Ph.D. degree in fields, particles and matter, *"mention très honorable avec félicitations"*. Subject of the thesis: "Neutrino astrophysics with the ANTARES telescope".

Occupation:

2008: Secondary School 7 Odintsovo, Russia, teacher.
Teaching: Information Science.

2008–now: Scobeltsyn Institute of Nuclear Physics, Moscow State University, researcher.
Field of studies: Neutrino detectors for astrophysics, the NEMO underwater telescope.

V. Kulikovskiy, *Neutrino Astrophysics with the ANTARES Telescope*,
Springer Theses, DOI 10.1007/978-3-319-20412-3

Duties: NEMO optical modules simulations, detector simulations for KM3NeT
Teaching: Course of neutrino physics for physics faculty students.

2009–2011: Fellowship, INFN Sezione di Genova.
Field of studies: Neutrino detectors for astrophysics, the ANTARES underwater telescope.
Duties: Data quality control, experimental setups for PMTs testing, measurement of neutrino emission from supernova with ANTARES, design of data filters to reduce the bioluminescence background noise in underwater neutrino detectors.

2011–2013: Dipartimento di Fisica, Università di Genova, Italy. Ph.D. student, IDAPP European doctorate program, co-doctorate at APC, Paris VII.
Field of studies: neutrino detectors for astrophysics, the ANTARES underwater telescope.
Duties: Data quality control, experimental test benches for PMTs, search for neutrino signal from the Fermi bubbles, FPGA programming for the KM3NeT optical module control logic board (CLB), investigation of the possibility of neutrino mass hierarchy discrimination with ORCA.

2014: Fellowship, LNS Catania.
Field of studies: Neutrino detectors for astrophysics, the KM3NeT and ORCA underwater telescopes.
Duties: FPGA programming for the KM3NeT CLBs, qualification of the Detection Optical Modules (DOMs) and PMTs. Simulation and the sensitivity estimation for the Dark Matter coming from the annihilation in the Sun and the Galactic Halo for the KM3NeT detector.
Teaching: Occasional lectures for the particle physics students in MSU and the University of Catania.

Skills

Language skills: Russian (native), English (fluent), Italian (fluent), French (advanced), German (basic).
Computer skills: Linux, C++, Python, Java, VHDL, SQL, ROOT, Geant4.
Hardware skills: CAMAC, VME, Xilinx FPGA, Optical networks, fast lasers, PMTs.

Participation in Schools and Conferences

1. 2010 40th Saas-Fee Course Astrophysics at Very-High Energies, Les Diablerets, Switzerland.
2. 2011 32nd ICRC Beijing, China, poster on “SN neutrino detection in the ANTARES neutrino telescope”.
3. 2011 The Neutrino Physics and Astrophysics, ISAPP School, Varenna, Italy, poster on “SN neutrino detection with the ANTARES detector”.
4. 2012-24/07/2012 CMB and the large scale structures, ISAPP School, La Palma, Spain, poster on “Search for neutrino emission from the Fermi Bubbles with ANTARES”.

5. 2012 NEUTRINO 2012 conference, Kyoto, Japan. Poster on "Search for neutrino emission from the Fermi bubbles with ANTARES".
6. 2013 33rd ICRC Rio De Janeiro, presentation on "A search for Neutrino Emission from the Fermi Bubbles with the ANTARES Telescope".
7. 2014 10th Rencontres du Vietnam, Quy Nhon, presentation, invited, "Search for diffuse cosmic neutrino fluxes with the ANTARES detector."

Publications with a Significant Impact

1. M. Taiuti et al. (31 authors). "A New Multianodic Large Area Photomultiplier to be used in Underwater Neutrino Detectors", Nucl. Instr. Meth. Phys. A 605 (2009), 293–300.
2. Creusot, et al. (4 authors) "PMT measurements in ANTARES", Nucl. Instr. Meth. Phys. A, 725 (2013) 144–147.
3. D. Franco, et al. (9 authors), "Mass hierarchy discrimination with atmospheric neutrinos in large volume ice/water Cherenkov detectors", JHEP, 04 (2013) 008.
4. The ANTARES collaboration (148 authors, corresponding author), "A Search for Neutrino Emission from the Fermi Bubbles with the ANTARES Telescope", EPJ C (2014) 74:2701.

5. 2014 NEUTRINO2014 conference, Kyoto, Japan: Poster on Search for neutrino emission from the Fermi Bubbles with ANTARES.
6. 2013 [illegible]: presentation on "A search for Neutrino Emission from the Fermi Bubbles with the ANTARES telescope".
7. 2014 [illegible] Vulcano, Italy: [illegible] presentation [illegible] ANTARES [illegible].

Publications [illegible] significant [illegible]

1. [illegible] A New Method [illegible]
2. [illegible] Phys. [illegible]
3. D. Fargion [illegible] large volume [illegible] telescopes [illegible]
4. The ANTARES collaboration (148 authors, [illegible]), A Search for Neutrino Emission from the Fermi Bubbles with the ANTARES Telescope, EPJ C (2014) 74:2701

Zeitfracht Medien GmbH
Ferdinand-Jühlke-Straße 7
99095 Erfurt, Deutschland
produktsicherheit@kolibri360.de